Philosophy in Neuroscience

Philosophy in Neuroscience

Edited by

JERZY STELMACH
BARTOSZ BROŻEK
ŁUKASZ KUREK

Copernicus Center
PRESS

Editing & Proofreading:
Aeddan Shaw
Piotr Godlewski

Cover design:
Mariusz Banachowicz

Cover image:
Andrzej Wróblewski, *Portret organiczny*, by kind permission
of the Andrzej Wróblewski Foundation

Layout:
Mirosław Krzyszkowski

Typesetting:
MELES-DESIGN

Publication supported by the John Templeton Foundation Grant
"The Limits of Scientific Explanation"

ISBN 978-83-7886-007-5

Kraków 2013

Copernicus Center PRESS

Publisher: Copernicus Center Press Sp. z o.o.,
pl. Szczepański 8, 31-011 Kraków,
tel/fax (+48) 12 430 63 00
e-mail: marketing@ccpress.pl
www.ccpress.pl

Printing and binding by *Colonel SA*

Preface

This volume brings together contributions which investigate problems at the intersection of philosophy and neuroscience. They are not, however, intended as an exercise in the philosophy *of* neuroscience: the structure of neuroscientific theories or the criteria of justification in neuroscience are not the primary subject of analysis. Instead, the papers collected here underscore the presence of philosophical problems within neuroscientific practice. In this, they follow the idea of philosophy *in* science proposed by Michael Heller. Heller believes that the sciences are filled with philosophically 'active' issues, and – to illustrate – he indicates that philosophical ideas have often influenced the development and evolution of scientific theories; that traditional philosophical concepts – such as space, time, mind or free will – are intertwined with empirical theories; and that the presuppositions of the empirical sciences are a legitimate, and highly interesting subject of philosophical analysis. Thus, by investigating philosophy *in* neuroscience, one not only engages in genuine philosophical work, but can also uncover what lies beneath the practice of neuroscience.

The first part of the collection, *Methodological Insights*, concentrates on the conceptual framework of philosophy *in* neuroscience. Bartosz Brożek in *Philosophy and Neuroscience. Three Modes of Interaction*, describes various ways of depicting the relationship between philosophical and neuroscientific theories. He argues that two such ways – isolationism and replacement – although quite popular in recent literature, rest upon a mistake as they both embrace a kind of epistemological or ontological foundationalism. He further posits that in order to grasp the relationship between philosophy and neuroscience, and to understand how both disciplines can enrich one another, one needs to dispense with foundationalism and acknowledge that both philosophical and neuroscientific argumentation is essentially non-foundational.

The goal of Robert Poczobut's *Mind, Emergence, and the Limits of Neuroscientific Explanations* is to determine the role of philosophy in the context of neuroscientific developments and to reveal the limits of neuroscientific explanation of mental phenomena. Poczobut underscores that neuroscience is a part of a broader research program of cognitive science. Furthermore, he believes that within the so-called 'philosophy in the context of neuroscience' one should distinguish between neurophilosophy (which applies neuroscientific findings to classical philosophical issues) and the philosophy of neuroscience (a specialized branch of philosophy of science). Poczobut also analyzes the important role of the concept of emergence, which enables one to take into account the relations between different levels of organization of cognitive phenomena. He concludes that although neuroscientific explanation provides invaluable insights at the neural level of organization, it is insufficient for a comprehensive account of mental phenomena.

The subsequent chapter *Theoretical Unification and the Neural Engineering Framework* by Marcin Miłkowski is an attempt to answer the question of whether the Neural Engineering Framework (NEF) is a possible unifying framework of explanation in cognitive neuroscience. According to Philip Kitcher, one can deal with the objections to the received view of scientific explanation (the covering-law account developed by Carl Hempel and Paul Oppenheim) in two distinct ways: by dispensing with the argument-like structure of the explanation and saving its causal character, or by accepting the argument-like structure and rejecting the concept of a law. After analyzing several counterexamples to Kitcher's unification account, Miłkowski argues that is not an adequate background for testing NEF's explanatory power. Miłkowski claims further that NEF exhibits the properties of frameworks that can serve as means of theoretical unification. On the basis of two case studies, one pertaining to the research on the navigational capabilities of rats and the other to the analysis of human performance in Wason selection tasks, Miłkowski claims that NEF offers a common methodology that unifies research on quite different phenomena.

The paper by Radosław Zyzik, *Biological Determinism and Its Enemies*, develops an account of biological determinism based on insights from neuroscience, behavioural genetics and neurogenetics. Zyzik analyzes three case studies: the famous neuroscientific experiment of Benjamin Libet (and its redesigns) concerning the neural basis of decision-making processes, the research on a single-gene defect of phenylketonuria which leads to severe retardation, and the neurogenetic research on *C. Elegans* – a model organism for the investigation of the interplay between genes, the nervous system and the environment. On the basis of his analysis, Zyzik argues that our knowledge about the biological factors determining behaviour is fragmentary at best, and the thesis of biological determinism, as proposed by many of its contemporary adherents, is ill-founded.

Wojciech Grygiel, in *Quantum Mechanics and Its Role in Cognitive Sciences: A Critical Survey*, reviews and critically assesses the applications of quantum mechanics in the explanation of neural and mental phenomena. The paper begins with an outline of the aspects of quantum mechanics presumably relevant for the research in cognitive science. Grygiel underlines, however, that due to the scale of single neurons, quantum effects may be extremely short-lasting to be of any relevance at the neuronal level. Next, the analysis of several proposals of the utilization of quantum mechanics in cognitive science is carried out: Roger Penrose and Stuart Hameroff's, John C. Eccles', Henry Stapp's and David Chalmers'. It is argued that at least the latter three accounts offer strong support for mind-body dualism. Furthermore, as Grygiel claims, in all of those proposals it is difficult to explain in detail how quantum effects are supposed to influence neural activity. Moreover, Stapp's and Chalmers' understanding of the relation between quantum mechanics and mental phenomena rests on some dubious philosophical assumptions.

The second part of the collection, *At the Border Between Neuroscience and Philosophy*, is devoted to more concrete issues arising at the intersection of both disciplines. Łukasz Kurek in *Philosophical Issues in the Neuroscience of Mindreading: The Case of Belief* proposes a philosophical account of belief developed in the context of the neuroscience of mindreading (i.e., the ability to predict, explain, and describe people's behaviour). He outlines how the concept of *belief* is understood in the so-called theory-theory of mindreading and the simulation theory of mind. Furthermore, he argues that on the theory-theory view belief is understood as involving complex and flexible cognitive processes, correlated with specific parts of the brain such as right temporo-parietal junction. On the other hand, within the simulation theory, belief-like states involve simple and inflexible cognitive processes, correlated with such parts of the brain as amygdala. Kurek claims that both these accounts are only partially correct and argues for the hybrid account of mindreading. According to this view, belief can be understood as a mental state utilized in high-level (complex and flexible) mindreading.

The following paper by Julia Stanek, *Self-Consciousness*: *From Philosophy to Neuroscience*, investigates some selected aspects of the problem of self-consciousness on the basis of several approaches: the philosophical, psychological, sociological, clinical and the neuroscientific. Stanek provides an overview of the conceptions of self-consciousness offered by such thinkers as Rene Descartes, John Locke, and Thomas Nagel. From the psychological perspective, the influential works of William James and Karl Jaspers are briefly outlined. Those considerations prepare the ground for a more detailed analysis of the research on self-consciousness carried out within neuroscience. Referring to several studies, Stanek argues that the phenomenon in question does not arise from a single neural process. Furthermore, she points out the conceptual and methodological difficulties connected with the neuroscience of self-consciousness.

In their paper *Self-Deception. Between Philosophy and Cognitive Neuroscience*, Mateusz Hohol and Piotr Urbańczyk indicate how the empirical investigations on self-deception are connected with the philosophical questions pertaining to this issue. They argue that the main point in the philosophical debate on self-deception is whether self-deception is intentional or not, and underscore some difficulties connected with both views. In the second part of the paper, three models of self-deception are proposed: the computational, the evolutionary, and the embodied. Hohol and Urbańczyk argue that intentionalist approaches to self-deception are more coherent with the embodied model of the phenomenon. However, the anti-intentionalist conceptualizations can still be useful in empirical investigations of the problem.

In *The Neurobiology of Depression* Jerzy Vetulani outlines the results of empirical research on depression. He points out the important factors correlated with this phenomenon, underlining the role of neurotransmitters (e.g. monoamine neurotransmitters), stress (including neonatal stress) and changes in brain structure (such as decrease in the volume of frontal cortex). Despite the growing body of knowledge concerning depression, Vetulani claims that the neurobiological processes underlying this disease are still poorly understood.

The subsequent paper, *The Moral Brain: Is There a Link Between Neurobiology and Morality?*, by Dominika Dudek, Rafał Jaeschke and Marcin Siwek, aims at identifying the neural underpinnings of morality. In the first part, they indicate the difficulties with the conceptualization of the main categories within morality, as well as the relative character of moral valence of action. After clarifying some key notions, Dudek, Jaeschke and Siwek make an attempt to answer the question of whether neuroscience can provide us with an explanation of morally valent behaviour. A survey of empirical research on this issue (such as the trolley problem experiments, neuropathologies, and the data derived from psychopathy) leads to the conclusion that, despite only fragmentary understanding of neural substrates of moral behaviour, one should acknowledge that the neurobiology of morality is no longer an abstraction.

The goal of Bram T. Heerebout and R. Hans Phaf's paper *Behavioral Benefits and Neural Costs in Simulated Evolution* is to substantiate the claim that time constraints imposed on the brain play an essential role in behaviour, and can explain actions that intuitively seem maladaptive. First, they argue that computational modelling methods provide insights which overcome shortcomings of theoretical evolutionary psychology. In the subsequent part of the paper, Heerebout and Phaf argue that evolutionary simulations based on these methods show that affective processes may be maladaptive, but on the whole they endow the organism with fitness-enhancing capabilities.

In the last paper of the volume, *On (Un)Happiness From the Viewpoint of Evolutionary Theory and Neuroscience*, Wojciech Załuski collects insights pertaining to the problem of happiness that come from philosophy, evolutionary theory and neuroscience. Załuski distinguishes between three types of questions related to happiness: the conceptual (what is happiness?), the empirical (what are the conditions of happiness?), and the axiological (can happiness be an ethical ideal?). He argues that evolutionary theory and neuroscience can provide us with insights only in regard to the empirical question, the other two questions remaining beyond the limits of scientific explanation.

The papers collected in this volume have been written within the research project *The Limits of Scientific Explanation*, carried out at the Copernicus Center for Interdisciplinary Studies in Kraków and sponsored by the John Templeton Foundation.

Jerzy Stelmach
Bartosz Brożek
Łukasz Kurek

Table of Contents

Part I

Methodological Insights

Bartosz Brożek
Philosophy and Neuroscience: Three Modes of Interaction 15

Robert Poczobut
Mind, Emergence, and the Limits of Neuroscientific Explanations 59

Marcin Miłkowski
Theoretical Unification and the Neural Engineering Framework 85

Radosław Zyzik
Biological Determinism and Its Enemies 113

Wojciech Grygiel
Quantum Mechanics and Its Role in Cognitive Sciences:
A Critical Survey 149

Part II

At the Border Between Neuroscience and Philosophy

Łukasz Kurek
Philosophical Issues in the Neuroscience of Mindreading:
The Case of Belief 175

Julia Stanek
Self-Consciousness: From Philosophy to Neuroscience 201

Mateusz Hohol, Piotr Urbańczyk
Self-Deception: Between Philosophy and Cognitive Neuroscience 219

Jerzy Vetulani
The Neurobiology of Depression 255

Dominika Dudek, Rafał Jaeschke, Marcin Siwek
The Moral Brain: Is There a Link Between Neurobiology and Morality? 279

Bram T. Heerebout, R. Hans Phaf
Behavioural Benefits and Neural Costs in Simulated Evolution 299

Wojciech Załuski
On (Un)Happiness from the Viewpoint of Evolutionary Theory and Neuroscience 331

Part I

Methodological Insights

Bartosz Brożek
Jagiellonian University
Copernicus Center for Interdisciplinary Studies

Philosophy and Neuroscience
Three Modes of Interaction[1]

Introduction

The relationship between philosophy and neuroscience[2] may be – and has been – accounted for in various ways. The first option is to claim that philosophy is firmly isolated from neuroscience – that they are mutually independent, or even that philosophical considerations are, in some way or another, *a priori* to neuroscientific endeavours. The second view is that of replacement: it may be argued that science in general – and neuroscience in relation to some philosophical issues – answers all the important questions of philosophy, but does

---

[1] This paper was written within the research grant "The Limits of Scientific Explanation" sponsored by the John Templeton Foundation. It is largely based on, and further develops the ideas expressed in, my previous publications: B. Brożek, *Philosophy in Neuroscience*, [in:] *Philosophy in Science*, eds. B. Brożek, J. Mączka, W. Grygiel, Copernicus Center Press, Kraków 2011, pp. 163–188; B. Brożek, A. Olszewski, *Logika zapętleń*, [in:] *Oblicza racjonalności*, eds. B. Brożek, J. Mączka, W. Grygiel, M. Hohol, Copernicus Center Press, Kraków 2011, pp. 33–50; and B. Brożek, *Rule-following. From Imitation to the Normative Mind*, Copernicus Center Press, Kraków 2013.

[2] The term "neuroscience" is understood very broadly here, and refers to, *inter alia*, behavioural neuroscience, cellular neuroscience, clinical neuroscience, cognitive neuroscience, computational neuroscience, cultural neuroscience, developmental neuroscience, molecular neuroscience, neuroimaging, neuroengineering, neuroinformatics, neurolinguistics, social neuroscience and systems neuroscience. However, I believe that my conclusions are applicable, *mutatis mutandis*, to any refinement of the term, i.e. to any reasonable rendering of 'neuroscience *sensu stricto*'.

so in a better way – not through mere speculation, but on the basis of solid empirical data. According to this account, scientific scrutiny has already replaced much of what used to be the domain of philosophical argument. I believe that both these stances are erroneous; however, it is interesting – and highly relevant – to ask, why this is the case. The analysis provided below suggests that both mistakes – of isolationism and of replacement – can be traced back to the same source: the assumption that either philosophy or neuroscience provides us with foundational, unshakable knowledge. I posit that only the rejection of this assumption – i.e., the appreciation of the fact that in philosophy, as well as in neuroscience, one has to do with non-foundational argumentation – opens the way for a truthful description of the interplay between the two disciplines and to the claim that they can *enrich* one another in many ways.

1. Isolation

The isolationist projects are based on the assumption that the findings of the natural sciences, neuroscience included, do not influence directly the practice of philosophy, as at least some aspects of philosophical reflection are independent of empirical facts investigated by scientists. The doctrine of isolation takes various shapes and forms, from some incarnations of the classical philosophy of Aristotle and Aquinas, through Kant's critical project and post-Kantian philosophies, such as phenomenology, to some versions of analytic philosophy. Let us have a look at two instructive examples.

The defenders of the contemporary versions of Thomism underscore the *autonomy* of philosophical thinking:

> The autonomy of Thomism boils down to the fact that its point of departure, as well as justification criteria, are independent of the truth of revelation as well as the findings of the natural sciences. The results of those disciplines can only (and often do) constitute the source of inspi-

> ration for new philosophical questions and determine new issues for metaphysical reflection. The maximalism (of Thomism) is connected to the fact that the goal of philosophizing is to uncover the fist and ultimate causes of the entire reality, including the cause of all causes – the Absolute, which makes the world intelligible and frees philosophical explanation from absurdity.[3]

Thus, the representatives of Thomism stress repeatedly that the autonomy of philosophy hangs together with its specific object and method: while the sciences consider only the so-called proximate causes of things, philosophy is capable of uncovering the ultimate causes of reality. Because of that, no empirical finding can falsify – or serve as a means for the rejection of – a philosophical theory. One should rather speak of two separate *planes of reflection*, the philosophical and the scientific, and if there is any relationship between them, it is the philosophical method that represents a higher, more profound mode of cognition.

This is an example of foundationalism in philosophy. Thomists believe that there exists only one true view of reality, captured by the Aristotelian-Thomistic conceptual scheme and penetrable by the Aristotelian-Thomistic method. All three dimensions of this foundationalism – the ontological, conceptual and methodological – prevent the findings of neuroscience from having any bearing on philosophical discourse: the sciences investigate only the manifestations of substances, utilize a method which cannot account for beings *qua* beings, and hence take advantage of a conceptual scheme which is not translatable into the metaphysical conceptual scheme and is inferior to it. The problem is that any kind of foundationalism in philosophy leads to daring consequences, when the relationship between philosophy and science is considered. As Michael Heller puts it:

---

[3] A. Maryniarczyk, *Tomizm*, [in:] *Powszechna encyklopedia filozofii*, http://www.ptta.pl/pef/.

> Today, after 300 years of the dynamic development [of the natural sciences], the employment of the strategy [of isolation] leads to two different kinds of danger. Firstly, some deep questions of obvious philosophical character (Did life originate from inanimate matter with no external factor at play? Is human brain only a perfect calculator?) may be rejected as no genuine philosophical issues (as they cannot be formulated within a given philosophical system). Secondly, artificial and highly confusing problems arise when one tries to speak of nature using a language which is inadequate for this purpose (i.e., a language of a certain philosophical system).[4]

What Heller stresses here is that faith in a philosophical system – in unshakable ontological or conceptual foundations – may easily lead to dispensing with real problems and to devoting time and effort to pseudo-problems. A closed, isolated philosophical system, one that provides answers to any questions, but only such that can be formulated within its conceptual framework, generates neither truth nor understanding, and hence becomes a caricature of what philosophical reflection should be.

More elusive, although equally destructive, is the isolationist stance defended by Bennett and Hacker in their celebrated book *Philosophical Foundations of Neuroscience.*[5] Bennett and Hacker insist, first, that one should clearly distinguish between two types of questions, conceptual and empirical:

> Distinguishing conceptual questions from empirical ones is of the first importance. (…) Conceptual questions antecede matters of truth and falsehood. They are questions concerning our forms of representation, not questions concerning the truth or falsehood of empirical statements. These forms are presupposed by true (and false) scientific

[4] M. Heller, *Nauki przyrodnicze a filozofia przyrody*, [in:] M. Heller, *Filozofia i Wszechświat*, Universitas, Kraków 2006, p. 28.

[5] M.R. Bennett, P.M.S. Hacker, *Philosophical Foundations of Neuroscience*, Wiley–Blackwell, Malden–Oxford 2003.

> statements and by correct (and incorrect) scientific theories. They determine not what is empirically true or false, but rather what does and what does not make sense. Hence conceptual questions are not amenable to scientific investigation and experimentation or to scientific theorizing. For the concepts and conceptual relationships in question are presupposed by any such investigations and theorizing.[6]

The final sentence of the quoted passage is of special interest. Bennett and Hacker claim that concepts are *a priori* to any scientific investigation. They believe further that the failure to notice this fact often leads to serious errors, and in particular the so-called mereological fallacy, common – as they stress – in contemporary cognitive neuroscience. It consists in referring to the brain or its parts with concepts which are correctly applicable only to a person as a whole. They observe:

> [talking] of the brain's perceiving, thinking, guessing or believing, or of one hemisphere of the brain's knowing things of which the other hemisphere is ignorant, is widespread among contemporary neuroscientists. This is sometimes defended as being no more than a trivial *façon de parler*. But that is quite mistaken. For the characteristic form of explanation in contemporary cognitive neuroscience consists in ascribing psychological attributes to the brain and its parts in order to explain the possession of psychological attributes and the exercise (and deficiencies in the exercise) of cognitive powers by human beings.[7]

One may ask whether the problem Bennett and Hacker identify is a real one. It may be argued, for example, that such claims as 'the brain thinks' or 'the right hemisphere is responsible for decision-making' are not to be taken literally. Some fundamental linguistic intuitions and a basic knowledge of language are enough to realize that

---

[6] M. Bennett, D. Dennett, P. Hacker, J. Searle, *Neuroscience and Philosophy: Brain, Mind, and Language*, Columbia University Press, New York 2007, p. 4.

[7] *Ibid.*, p. 7.

such a utilization of the words 'think' or 'decide' is metaphorical or analogical. Bennett and Hacker are fully aware of this strategy to defend the existing neuroscientific idiom. They elaborate it further claiming that the strategy may be used in four different ways. First, one may insist that psychological concepts used by neuroscientists, e.g. 'to think', have a different, derivative meaning to the meaning of the terms in the ordinary language. Second, the 'neuroscientific meaning' of a term may be analogical or constitute some other extension of the meaning of the corresponding ordinary language concept. Third, it may also be treated as a homonym: 'to think' or 'to decide' in a description of brain functions may have an altogether different meaning than the corresponding ordinary language terms. Finally, such concepts in neuroscience may be treated as metaphorical expressions.

Bennett and Hacker believe, however, that the above described strategies fail. They put forward a number of arguments to back this claim; the most important, and which is applicable to all four strategies, is the following. According to Bennett and Hacker, there exists a criterion that suffices to show that the use of psychological terms in neurobiology is neither a case of taking advantage of derivative meaning, nor of analogical, homonymous or metaphorical. The criterion in question is the analysis of *conclusions* that the neuroscientists draw from the claims such as 'the brain thinks'. Let us have a look at an example. Colin Blakemore notes:

> We seem driven to say that such neurons [as they respond in a highly specific manner to, e.g., line orientation] have knowledge. They have intelligence, for they are able to estimate the probability of outside events – events that are important to the animal in question. And the brain gains its knowledge by a process analogous to the inductive reasoning of the classical scientific method. Neurons present arguments to the brain based on the specific features that they detect, arguments on which the brain constructs its hypothesis of perception.[8]

[8] *Ibid.*, p. 16.

In this passage Blakemore claims that 'neurons possess knowledge'. However, he does not end here; on the basis of this observation he constructs a complex conception of the interaction between neurons and the brain, which utilizes almost exclusively psychological terminology ('intelligence', 'inductive reasoning', 'construction of a perceptual hypothesis'). But why can't we consider this *entire* passage as one complex metaphor? Blakemore, in a different context, observes:

> Faced with such overwhelming evidence for topographic patterns of activity in the brain it is hardly surprising that neurophysiologists and neuroanatomists have come to speak of the brain having maps, which are thought to play an essential part in the representation and interpretation of the world by the brain, just as the maps of an atlas do for the reader of them. (...) But is there a danger in the metaphorical use of such terms as 'language', 'grammar', and 'map' to describe the properties of the brain? (...) I cannot believe that any neurophysiologist believes that there is a ghostly cartographer browsing through the cerebral atlas. Nor do I think that the employment of common language words (such as map, representation, code, information and even language) is a conceptual blunder of the kind [imagined]. Such metaphorical imagery is a mixture of empirical description, poetic license and inadequate vocabulary.[9]

Here, however, Bennett and Hacker launch their counter-attack. They ask how one should understand such claims as 'the brain interprets the world'. They suggest that Blakemore's use of 'metaphorical' expressions such as 'a map' leads directly to the utilization of inadequate terminology in the entire argumentation. This shows, as they believe, that there is no metaphor here; rather, Blakemore commits the mereological fallacy. In addition, they observe:

---

[9] *Ibid.*, p. 32.

> whatever sense we can give to Blakemore's claim that 'brain-maps' (which are not actually maps) play an essential part in the brain's 'representation and interpretation of the world', it cannot be 'just as the maps of an atlas do for the reader of them'. For a map is a pictorial representation, made in accordance with conventions of mapping and rules of projection. Someone who can read an atlas must know and understand these conventions, and read off, from the maps, the features of what is represented. But the 'maps' in the brain are not maps, in this sense, at all. The brain is not akin to the reader of a map, since it cannot be said to know any conventions of representations or methods of projection or to read anything off the topographical arrangement of firing cells in accordance with a set of conventions. For the cells are not arranged in accordance with conventions at all, and the correlation between their firing and features of the perceptual field is not a conventional but a causal one.[10]

Bennett and Hacker's position is that the evidence that neuroscientists commit the mereological fallacy does not lie in the fact that, on occasion, they use 'inadequate' psychological terms to describe the functioning of the brain, which may easily count as taking advantage of analogy, metaphor, homonym or using a concept with a derivative meaning. The mereological fallacy results when neuroscientists transfer entire complexes of concepts from 'psychological discourse' to the 'neuroscientific' one, and – on the basis of such inadequate attributions – they draw conclusions.

Is Bennett and Hacker's argument tenable? I believe not, and the reason is analogous to the case of the Thomistic isolationist project: their foundationalism. There are two interpretations of Bennett and Hacker's foundationalism. The stronger interpretation, attributed to them by John Searle, is that they believe natural language to determine the only acceptable ontology. Searle says that they commit a fallacy:

---

[10] *Ibid.*, p. 33.

> the fallacy, in short, is one of confusing the rules for using the words with the ontology. Just as old-time behaviorism confused the evidence for mental states with the ontology of the mental states, so this Wittgensteinian criterial behaviorism construes the grounds for making the attribution with the fact that is attributed. It is a fallacy to say that the conditions for the successful operation of the language game are conditions for the existence of the phenomena in question.[11]

This reading finds some textual evidence. Interestingly, while elaborating the doctrine of the mereological fallacy, Bennett and Hacker quote Aristotle as one of those who first condemned this erroneous mode of thinking. He observed that "to say that the soul is angry is as if one remarked that the soul weaves or builds, for it is surely better not to say that the soul pities, learns or thinks, but that a man does these with his soul". One needs to remember, however, there is a certain metaphysical view standing behind his claim. Aristotle's metaphysics is essentialist: he believes that every entity belongs to some natural category, one determined by the entity's essence (form); moreover, he believes that the essences may be captured by the so-called essential definitions.[12] Thus, the incorrect or metaphorical use of words is not a mere mistake – it is an error that may effectively ruin our attempts to construct the foundations of knowledge, captured by the essential definitions. This doctrine is, of course, far from actual scientific practice. The history of science shows clearly that no such foundations should be assumed as they are most likely to hinder scientific progress. But if so, the same holds for Bennett and Hacker's view: if they indeed believe that the conceptual scheme of the ordinary language determines 'the only' ontology, their conception is hopelessly flawed.

It is also possible to read Bennett and Hacker in a more moderate manner; this weaker interpretation is that they only underscore that

---

[11] *Ibid.*, p. 105.

[12] See K. Popper, *The Open Society and Its Enemies*, vol. II, Princeton University Press, Princeton 1966, p. 30.

the conceptual scheme which constitutes the framework for the ordinary language does not determine any unique ontology, but nevertheless is independent of any scientific practice, in the sense that in order to communicate any scientific discovery one needs to employ concepts according to some pre-existing criteria. If one does not do so, one risks following wrong paths and uttering nonsensical statements: the incorrect use of language can lead us astray. The conceptual scheme of ordinary language constitutes, at the very least, the foundation for communicating scientific theories.

This incarnation of foundationalism is as equally troublesome as its Aristotelian-Thomistic predecessor. Firstly – and less importantly – Bennett and Hacker are mistaken when they claim that an excessive use of metaphors, and in particular – clusters of metaphors – is destructive for any neuroscientific endeavour. Certainly, it may lead to blind alleys yet there is little danger that the consequences of such a way of expression will be daring. The reason is that neuroscience, as with any other science, has some in-built corrective mechanisms that ultimately help us to distinguish progress and fruitful hypotheses from mere mistakes and useless conjecture. That this mechanism is present is evident once one considers the recent successes of neuroscience. A science which overuses metaphors and leads to no serious predictions or explanations is simply a bad science; the mere fact of committing or omitting the mereological fallacy is of no significance here.

Secondly, we should consider the bigger picture, which is encapsulated in Bennett and Hacker's claim that the conceptual scheme of ordinary language is *a priori* relative to scientific practice. It is particularly troublesome with respect to neuroscience. It must be realized that the psychological idiom, characteristic of the ordinary language, is not only shaped by our inner experience, but also by the theories developed throughout history which aimed at conceptually capturing mental phenomena. The problem is that the conceptual scheme of ordinary language is characterized by some inertia: it takes a considerable amount of time for current scientific conceptions to 'infiltrate' our ordinary conceptual scheme. It is safe, therefore, to assume that today's ordinary language 'embraces' some psychological theories of yesterday, or better even: a blend of those theories and common-sense ideas, often referred to as the folk psychology. Now, to say that ordinary concepts are *a priori* relative to neuroscience amounts to saying that folk psychology is *a priori* to neuroscientific theories, which is outright nonsense: it is one of the main goals of the contemporary neuroscience, one that it fulfils vigorously and with much success, to revise our old, common-sense psychological notions.

This clearly shows that Bennett and Hacker's conceptual foundationalism is faithful neither to the mechanisms of scientific practice, nor the way our conceptual schemes evolve: they are never final, or independent of the theories we develop. This point is quite general and pertains to any foundational philosophical project: the sources of philosophical reflection are always, at least partially, based on *some* scientific conceptions, although often on outdated ones. In the case of Thomism, the Aristotelian view of the world – or the Aristotelian science – constitutes the foundations of the conceptual scheme. Similarly, in the case of those philosophies that find confirmations or disconfirmations in the workings of the ordinary language, it is the knowledge encapsulated there (e.g., a kind of folk psychology being a blend of the common sense observations and some old psychological theories) that ultimately determines the philosophical doctrines of the followers of Austin and Strawson. In other foundational projects, such as phenomenology, the scientific knowledge internalized by any given person crucially shapes this person's experiences, and so her philosophical views. All in all, there is no source of philosophical knowledge that would be independent of some kind of science, and the key point is that this 'hidden science' may be at odds with what the contemporary science has to say. The splendid isolation of philosophers is an illusion: there is no escape from the confrontation with the barbarians from the other side of the Channel.

2. Replacement

The second stance towards the relationship between philosophy and neuroscience is one that sees an inherent conflict between the two. This seems to be mainly the perspective adopted by neuroscientists who believe that the findings of neuroscience resolve philosophical problems and, what follows, scientific method and scientific theories should *replace* philosophical reflection. Thus, there is a conflict here, but one in which there may be only one survivor. The problem is that – similarly to the case of isolationism – those who believe in the replacement strategy base their convictions on a foundational view of knowledge.

In order to illustrate how neuroscientists tend to approach philosophical problems, let us have a look at the neuroscience of mathematics. This field of knowledge has made impressive progress during the past twenty years. Although there is no single, commonly accepted theory of the neural basis of mathematical skills, a general picture that emerges from the findings of neuroscience may be deemed the 3E view of mathematics; mathematics which is embrained, embodied and embedded. The general elements of this view are the following. Firstly, the current research implicates two separate brain systems as responsible for the basic numerical capacities: the object tracking system (OTS) and the approximate number system (ANS). OTS is a system that enables the tracking of multiple individuals (up to 3 or 4). It is based on the principles of cohesion (moving objects are recognized as bounded wholes), continuity (objects move on unobstructed paths) and contact (objects do not interact at a distance).[13] The existence of the OTS system is confirmed by a number of tests, including visual short-term memory tasks, multiple-objects tracking tasks, or enumeration tasks. The last kind of test confirms human

[13] M. Piazza, *Neurocognitive Start-Up Tools for Symbolic Number Representations*, [in:] *Space, Time and Number in the Brain*, eds. S. Dehaene, E. Brannon, Academic Press, London 2011, p. 270.

ability of *subitizing*, i.e. of an instant and highly accurate determination of a number of objects in small collections (3–4), even presented very briefly.[14] Furthermore, it is speculated that the posterior parietal and occipital regions of the brain play the crucial role in the performance of such tasks, which suggests that these regions are the location of OTS.[15]

ANS, on the other hand, is a system for representing the approximate number of items in sets. It works according to Weber's Law: the threshold of discrimination between two stimuli increases linearly with stimulus intensity. In the case of ANS, the Weber fraction, or the smallest variation to a quantity that can be readily perceived, changes over human development. For newborns it is 1:3, for 6-month-old babies it is 1:2, for 1-year-old children it is 2:3, for 4-year-olds it is 3:4, for 7-year-olds it is 4:5, while for 20-year-olds it is 7:8. It means that a newborn can discriminate between 1 and 3, or 2 and 6, or 10 and 30, but not 1 and 2, 2 and 5, or 10 and 27. Four-year-old children can tell that there is a difference in numerosity between sets consisting of 6 and 8 or 12 and 16 elements, but not 7 and 8 or 12 and 15. Adults' ANS system is even more 'sensitive': they can discriminate (without counting) between sets consisting of 14 and 16 elements or 70 and 80 elements, but not 70 and 78 elements. It is quite well established that ANS is located in the mid-intreparietal sulcus.[16]

OTS and ANS constitute innate or 'embrained' arithmetic skills and are clearly quite limited. The question is, how people move beyond these limited inborn abilities and acquire 'full-blooded' mathematical skills. There are a number of hypotheses explaining this process, but most of them point out to the key role of language in both phylogenetic and ontogenetic development of arithmetical ability. The hypothesis that the development of counting skills is conditioned and mediated by the acquisition of language is supported

[14] *Ibid.*, p. 271.
[15] *Ibid.*, p. 270.
[16] *Ibid.*, p. 268–269.

by the following evidence. First, both children and adults in remote cultures, whose languages have no words for numbers, when dealing with numbers larger than three only recognize their equivalence approximately. Second, an interesting line of evidence comes from the study of the mathematical abilities of deaf people. Deaf persons living in numerate cultures but not exposed to the deaf community use a gestural system called homesign; they use fingers to communicate numbers, but only with approximate accuracy. Similarly, they perform matching tasks with approximate accuracy.[17] Moreover, language continues to play an important role in mathematical cognition even after the mathematical skills are mastered. For instance, educated adults who suffer language impairments have problems with exact, but not approximate numerical reasoning. Similarly, when doing exact (but not approximate!) tasks, adults spend more time with numbers that are difficult to pronounce, even if they are presented in Arabic notation. But "if language merely scaffolded the acquisition of natural number concepts and abilities, and then was replaceable by other symbol systems, one would not expect adults to translate Arabic symbols into words for purposes of exact computation".[18] Finally, bilingual adults who are taught some new mathematical facts in one of their languages have difficulties in the smooth production of exact number facts in the other language.[19]

These observations lead to an important question: if the acquisition of language plays such a crucial role in the ontogenetic development of mathematical skills, what exactly is the language-dependent mechanism that enables to move beyond embrained mathematics? One of the most intriguing, albeit controversial, answers to this question was formulated Lakoff and Núñez, within the so-called 'embodied mind' paradigm. The very idea of embodiment boils down to the thesis that human mind and human cognition are decisively shaped

[17] E.S. Spelke, *Natural Number and Natural Geometry*, [in:] *Space, Time and Number in the Brain*, eds. S. Dehaene, E. Brannon, Academic Press, London 2011, p. 306–307.
[18] *Ibid.*, p. 307.
[19] *Ibid.*, p. 307.

by the experiences of our bodies. This is a vague claim that only underscores the rejection of other paradigms, such as Cartesian mind--body dualism or computationalism (the rough idea that human brain is hardware, and the mind is software implemented in the brain). However, other claims of the representatives of the embodied mind approach are more informative. They believe that the human mind is a powerful conceptual system shaped by our bodies' experiences during their interactions with the environment. The most basic mental concepts or schemas, probably derived from the neural motor-control programs, express spatial relations (such as the Source – Path – Goal schema). Since such "image schemas are conceptual in nature, they can form complex composites. For example, the word 'into' has a meaning – the *Into* schema – that is the composite of an *In* schema and *To* schema".[20] Furthermore, this mental machinery is capable of producing abstract concepts with the use of concrete ones through the use of metaphors. In the 'embodied paradigm' metaphors are understood as the means for "understanding and experiencing one kind of thing in terms of another".[21] And so, importance is conceptualized in terms of size ("This is a big issue", "It's a small issue; we can ignore it"), difficulties are conceptualized as burdens ("He is overburdened", "I've got a light load this semester"), etc.[22] "Each such conceptual metaphor has the same structure. Each is a unidirectional mapping from entities in one conceptual domain to corresponding entities in another conceptual domain. As such, conceptual metaphors are part of our system of thought. Their primary function is to allow us to reason about relatively abstract domains using the inferential structure of relatively concrete domains".[23]

Lakoff and Núñez claim further that it is the mechanism of conceptual metaphorization that enables the construction of complex and

[20] G. Lakoff, R. Núñez, *Where Mathematics Comes From*, Basic Books, New York 2000, p. 39.
[21] *Ibid.*, p. 5.
[22] *Ibid.*, p. 41.
[23] *Ibid.*, p. 42.

precise mathematical concepts. In the case or arithmetic, they postulate the existence of four basic or grounding metaphors: the Arithmetic as Object Collection (where the source domain concept of collections of objects of the same size is mapped to the concept of numbers, the size of the collection is mapped to the size of the number, the smallest collection is mapped to the concept of the unit, while putting collections together is mapped to the process of addition); the Arithmetic as Object Construction (where the source domain concept of objects consisting of ultimate parts of unit size is mapped to the concept of numbers or the act of object construction is mapped to the concept of arithmetic operations); the Measuring Stick (where physical segments are understood as numbers, the basic physical segment as one, and the length of the physical segment as the size of the number); and the Arithmetic as Motion Along a Path (where the act of moving along the path is understood as representing mathematical operations, point-locations on the path are understood as numbers, etc.).[24]

Lakoff and Núñez claim that those four grounding metaphors give rise to the development of basic arithmetic. One begins with innate capacities to 'deal' with small numbers (up to 4). In addition, one has primary experiences with object collections, object construction, physical segmentation and moving along a path. "In functioning in the world, each of those primary experiences is correlated with subitizing, innate arithmetic, and simple counting".[25] Those two domains are combined through the four metaphors in such a way that the primary experiences become sources of the metaphors and the domain of numbers is the target of the metaphors. "The next step is the conflation among the primary experiences: object construction always involves object collection. Placing physical segments end to end is a form of object construction (…). From a neural perspective, [such conflations] involve co-activations of those brain areas that characterize each of the experiences. (…) As a consequence, an isomorphic

[24] *Ibid.*
[25] *Ibid.*, p. 93.

structure emerges across the source domains (...), which is independent of numbers themselves and lends stability to arithmetic".[26] The ability of subitizing, found in all normal human beings, leads to precise and stable results regarding small numbers; when extended with the four grounding metaphors, the precision and stability extends to all natural numbers. Finally, "the laws of arithmetic (commutativity, associativity and distributivity) emerge first as properties of the four source domains, then as properties of numbers *via* those metaphors, since the metaphors are inference-preserving conceptual mappings".[27]

If this picture is roughly correct, mathematics – although based on some simple innate or embrained skills – is also embodied and embedded. It is largely derived – through the process of metaphorization – from concrete concepts shaped by our bodies' interactions with the environment, as well as sustained and further developed through the interactions with others by imitating their behaviour. The embedding of mathematical practice in social interactions contributes to the stability of mathematical knowledge.

There are many philosophical problems connected to mathematics. Two of the most fundamental are the problem of the necessity in mathematics, and the problem of the mathematicity of the universe. Let us begin with the former. There is a dimension of mathematical and logical research that traditionally poses a challenge to any naturalistic accounts of the ontology of mathematical or logical objects. It is well captured in the following observation by Jan Łukasiewicz:

> Whenever I deal with the smallest logical problems, I always have the feeling that I am facing some powerful, incredibly coherent and enormously resistant structure. I cannot make any changes within it, I create nothing, but working hard I uncover new details, gaining eternal truths.[28]

---

[26] *Ibid.*, p. 95–96.
[27] *Ibid.*, p. 96.
[28] J. Łukasiewicz, *W obronie logistyki. Myśl katolicka wobec logiki współczesnej*, "Studia Gnesnensia" 1937, vol. 15, p. 14.

Such views as the one expressed by Łukasiewicz give rise to the development of mathematical Platonism (realism), a view that "mathematics is the scientific study of objectively existing mathematical entities just as physics is the study of physical entities. The statements of mathematics are true or false depending on the properties of those entities, independent of our ability, or lack thereof, to determine which".[29]

There are many forms of mathematical Platonism. In particular, one should distinguish between ontological Platonism (a view pertaining to the existence of mathematical objects) and semantic Platonism (an epistemological view that mathematical statements are true or false). Ontological Platonism is a stronger theory – it implies the semantic one, but the opposite implication does not hold. Thus, in what follows I shall concentrate on the stronger claim. Arguably, ontological Platonism in mathematics, though it comes in various incarnations, embraces the following three theses:

> (The existence thesis) Mathematical objects (or structures) exist.
> (The abstractness thesis) Mathematical objects are abstract, non-spatio-temporal entities.
> (The independence thesis) Mathematical objects are independent of any rational or irrational activities of the human mind. In particular, mathematical objects are not our constructions.[30]

The key question is how the above formulated theses are justified. With no pretence to comprehensiveness, I posit that there are three kinds of arguments backing mathematical Platonism in its ontological version. The first one is the *semantic argument*, well captured by Balaguer, but formulated earlier by Frege[31]:

---

[29] P. Maddy, *Realism in Mathematics*, Oxford University Press, Oxford 1990, p. 21.
[30] Cf. Ø. Linnebo, *Platonism in the Philosophy of Mathematics*, [in:] *The Stanford Encyclopedia of Philosophy* (Fall 2011 Edition), ed. E.N. Zalta, URL = <http://plato.stanford.edu/archives/fall2011/entries/ platonism-mathematics/>.
[31] Cf. *Ibid.*

(a) Mathematical propositions are true.
(b) Mathematical propositions should be taken at their face value. In other words, there is no reason to believe that mathematical propositions, as they appear, are not what they really are, or that there is a deep structure of mathematical propositions which differs from their surface structure, of what they seem at their face.
(c) By Quine's criterion, we are ontologically committed to the existence of objects which are values of the variables in the propositions we consider true.
(d) We are ontologically committed to the existence of mathematical objects.
(e) Therefore, there are such things as mathematical objects, and our theories provide true descriptions of these things. In other words, mathematical Platonism is true.

The second argument defending mathematical Platonism is *the indispensability argument*, or the Quine/Putnam argument. Muddy summarizes it in the following way: "we are committed to the existence of mathematical objects because they are indispensable to our best theory of the world and we accept that theory".[32] And in Putnam's own words: "mathematics and physics are integrated in such a way that it is not possible to be a realist with respect to physical theory and a nominalist with respect to mathematical theory".[33] A reconstruction of this argument may appear as follows:

(a) By Quine's criterion, we are committed to the existence of objects which our best physical theories speak of.
(b) Our best physical theories are expressed with the use of the language of mathematics.

[32] P. Maddy, *Realism…*, *op. cit.*, p. 30.
[33] H. Putnam, *What is mathematical truth?*, [in:] H. Putnam, *Mathematics, Matter and Method*, Cambridge University Press, Cambridge 1979, p. 74.

(c) Therefore, we are committed to the existence of mathematical objects.
(d) When one is a realist with respect to physical theories, one must also be a realist with respect to mathematics.
(e) Therefore, mathematical Platonism is true.

Finally, Gödel's *intuition-based argument* may be reconstructed in the following way:

(a) The most elementary axioms of set theory are obvious; as Gödel puts it, they "force themselves upon us as being true"[34].
(b) In order to explain (a), one needs to posit the existence of mathematical intuition, a faculty analogous to the sense of perception in the physical sciences.
(c) Not all mathematical objects are intuitable; but our belief in the 'unobservable mathematical facts' is justified by the consequences they bring in the sphere controllable by intuition and through their connections to already established mathematical truths. As Gödel says, "even disregarding the [intuitiveness] of some new axiom, and even in case it has no [intuitiveness] at all, a probable decision about its truth is possible also in another way, namely, inductively by studying its 'success' (…). There might exist axioms so abundant in their verifiable consequences, shedding so much light upon a whole field, and yielding such powerful methods for solving problems (...) that, no matter whether or not they are [intuitive], they would have to be accepted at least in the same sense as any well-established physical theory".[35]

Let us consider now, whether the 3E account of mathematics sketched above has any bearing on the arguments favouring mathe-

[34] K. Gödel, *What is Cantor's continuum problem?*, [in:] *Philosophy of Mathematics*, eds. P. Benacerraf, H. Putnam, Cambridge University Press, Cambridge 1983, p. 484.
[35] *Ibid.*, p. 477.

matical Platonism. Lakoff and Núñez believe that the conception of the embodied mathematics puts mathematical Platonism to eternal rest. For them, mathematical Platonism is "the romance of mathematics", a "story that many people *want* to be true"[36]; a story that mathematical objects are real, and mathematical truth is universal, absolute, and certain. They succinctly reject this view:

> The only access that human beings have to any mathematics at all, either transcendent or otherwise, is through concepts in our minds that are shaped by our bodies and brains and realized physically in our neural systems. For human beings – or any other embodied beings – mathematics *is* embodied mathematics. The only mathematics we can know is mathematics that our bodies and brains allow us to know. For this reason, the *theory of embodied mathematics* (…) is anything but innocuous. As a theory of the only mathematics we know or can know, it is a theory of what mathematics *is* – what it really is![37]

As I read them, Lakoff and Núñez emphasise two things. First, they put forward an epistemological claim that we have no cognitive access to independent abstract objects, since the only way of practicing mathematics is through the concepts "shaped by our bodies and brains". The problem of the cognition of abstract objects has been a subject of controversy since the beginnings of philosophy. Painting with a broad brush, one may claim that two solutions have been defended in this context, both already present in Plato: that there exists a rational intuition enabling us to contemplate abstract objects or that our access to the abstract sphere is discursive, mediated by language. Lakoff and Núñez seem to consider only the first option, and dismiss it on the basis of the recent findings in neuroscience.

Second, they seem to embrace a version of Quine's criterion: we are committed to the existence of only those things which our best

---

[36] Cf. G. Lakoff, R. Núñez, *Where…*, *op. cit.*

[37] *Ibid.*, p. 346.

scientific theories speak of. They add that the best – or rather: the only – theory of mathematical cognition we have is the theory of embodied mathematics, and since it does not speak of independent abstract objects, we have no grounds for postulating their existence. The problem is that Quine's criterion – applied to other theories, not necessarily accounting for the nature of mathematics, e.g. to our best physical theories – brings a different outcome: that we are indeed committed to the existence of abstract mathematical objects.

A similar line of argument is developed by Stanislas Deheane. Here is a longer passage that encapsulates his view well:

> For an epistemologist, a neurobiologist, or a neuropsychologist, the Platonist position seems hard to defend — as unacceptable, in fact, as Cartesian dualism is unacceptable as a scientific theory of the brain. Just as the dualist hypothesis faces insurmountable difficulties in explaining how an immaterial soul can interact with a physical body, Platonism leaves in the dark how a mathematician in the flesh could ever explore the abstract realm of mathematical objects. If these objects are real but immaterial, in what extrasensory ways does a mathematician perceive them? This objection seems fatal to the Platonist view of mathematics. Even if mathematicians' introspection convinces them of the tangible reality of the objects they study, this feeling cannot be more than an illusion. Presumably, one can become a mathematical genius only if one has an outstanding capacity for forming vivid mental representations of abstract mathematical concepts — mental images that soon turn into an illusion, eclipsing the human origins of mathematical objects and endowing them with the semblance of an independent existence.[38]

Thus, Deheane also stresses the epistemological point: that we have no, and cannot have any, cognitive access to the realm of abstract objects. However, also in this case, the epistemological point has de-

---

[38] S. Deheane, *The Number Sense*, 2nd edition, Oxford University Press, Oxford 2011, p. 225.

vastating consequences for mathematical Platonism: if we have no access to the abstract sphere, there are no grounds for postulating its existence. It is clearly visible that Deheane, similarly to Lakoff and Núñez, does not even consider that our grasping of abstract objects may be enabled by language.

It seems, moreover, that Lakoff and Núñez's, as well as Deheane's critiques, do not defeat any of the three arguments in favour of mathematical Platonism described above. In order to defeat the semantic argument one would have either to show that mathematical propositions cannot be ascribed truth or falsehood; or to reject the idea that mathematical propositions have no 'deep structure': that they are what they seem at their face; or to reject Quine's criterion of ontological commitment. Neither Lakoff and Núñez, nor Deheane, do so. Also, they fail to address the indispensability argument. To do so, they would need either to reject Quine's criterion; or the thesis that mathematical physics is our best theory of the world; or the realist stance towards physical theories. Finally, the intuition-based argument seems the easiest to attack from the point of view of the neuroscience of mathematics. As we have seen above, human 'intuitive' mathematical capacities are substantially limited. However, Gödel – the proponent of the intuition-based argument – does not claim that our intuition is a faculty that gives us access to the entire world of mathematical structures. His thesis is that intuition is the source of certainty in relation to relatively simple mathematical structures and relations; more complicated mathematical propositions are evaluated as true because they are justified by commonly accepted mathematical methods and have consequences controllable at the intuitive level. Of course, Lakoff, Núñez and Deheane may claim that the intuition Gödel speaks of is not an intuition of *abstract* objects; it is rather the capacity to use abstract mathematical concepts, which are ultimately shaped by the experiences of our bodies. But this criticism can be softened by a modification of Gödel's argument: instead of speaking of intuition, one can simply speak of mathematical experience, even conceived of in terms of Lakoff and Núñez's of of Deheane's theory. The crux of

Gödel's thesis, or so I argue, lies somewhere else: mathematical Platonism is true, because "there exist axioms so abundant in their verifiable consequences, shedding so much light upon a whole field, and yielding such powerful methods for solving problems that, no matter whether or not they are [a subject of direct experience], they would have to be accepted at least in the same sense as any well-established physical theory". Gödel points out to something important here: the full power of our abstract conceptions, which lie beyond any intuitive or 'direct' experience, is clearly visible in the consequences they produce within the sphere controllable by experience, as well as in the coherence they bring to entire areas of mathematics and the heuristic role they play in solving mathematical problems. It is reasonable, therefore, to assume that those highly abstract concepts describe some independently existing structures rather than claim that they are just 'metaphorizations' of more concrete concepts. The mathematics we can somehow experience directly is only the tip of the iceberg: and when Lakoff and Núñez, as well as Deheane believe that the rest of the iceberg is only an illusion, Gödel seems to claim that it is a rock-hard, even if abstract, reality.

All this is not to say that the three arguments supporting mathematical Platonism are irrefutable or incontestable: the heated debates in the philosophy of mathematics during the last century are the evidence to the contrary. However, Lakoff, Núñez and Deheane failed to provide a persuasive case against mathematical Platonism. Moreover, along the way, they have themselves accepted some philosophical assumptions, such as that the only access to the abstract objects is through some kind of rational intuition; or Quine's existence criterion; or realism in relation to biological theories.

The second problem I would like to address is that of the mathematicity of the universe. Michael Heller introduces the concept in the following words:

> In the investigation of the physical world one method has proved particularly efficient: the method of mathematical modeling coupled

> with experimentation (to simplify, in what follows I shall speak of the mathematical method). The advances in physics, since it has adopted the mathematical method, have been so enormous that they can hardly be compared to the progress in any other area of human cognitive activity. This incontestable fact helps to make my hypothesis more precise: the world should be ascribed a feature thanks to which it can be efficiently investigated with the use of the mathematical method. Thus the world has a rationality of a certain kind – a mathematical one. It is in this sense that I shall speak of the mathematicity of the universe.[39]

According to Heller, to say that the world is mathematical is equivalent to the claim that it possesses a feature which makes mathematical method efficient. In the quoted passage, Heller hints at one of the aspects in which the mathematicity of the world should be understood: the efficiency thesis. It says that the mathematicity of the universe is evident once one considers the enormous success of the mathematical method over the last 300 years. The success cannot be a matter of coincidence, as the efficiency of mathematics in uncovering the laws of nature seems 'unreasonable'.[40] The argument pertaining to the 'unreasonable effectiveness of mathematics' is not trivial. As Eugene Wigner observes:

> It is true, of course, that physics chooses certain mathematical concepts for the formulation of the laws of nature, and surely only a fraction of all mathematical concepts is used in physics. It is true also that the concepts which were chosen were not selected arbitrarily from a listing of mathematical terms but were developed, in many if not most cases, independently by the physicists and recognized then as having been conceived before by the mathematicians. It is not true, however, as is so often stated, that this had to happen because

[39] M. Heller, *Czy świat jest matematyczny?*, [in:] M. Heller, *Filozofia i wszechświat*, Universitas, Kraków 2006, p. 48.

[40] Cf. E. Winger, *The Unreasonable Effectiveness of Mathematics in the Natural Sciences*, "Communications on Pure and Applied Mathematics" 1960, vol. 13(1), pp. 1–14.

> mathematics uses the simplest possible concepts and these were bound to occur in any formalism. [Moreover], it is important to point out that the mathematical formulation of the physicist's often crude experience leads in an uncanny number of cases to an amazingly accurate description of a large class of phenomena. This shows that the mathematical language has more to commend it than being the only language which we can speak; it shows that it is, in a very real sense, the correct language.[41]

There are some phenomena connected to the use of the mathematical method that leads to the conclusion that it is some feature of the world that must be responsible for the method's successes. It is often the case that mathematical equations describing some aspects of the universe 'know more' than their creators. The standard story in this context is that of Einstein's cosmological constant. When Einstein formulated his cosmological equations on the basis of the newly discovered general relativity theory, he realized that they imply a dynamic, expanding universe. In order to 'stop' the expansion, he introduced the cosmological constant. It quickly proved, however, that Einstein was 'wrong' and his equations were 'right': the expansion of the universe is a fact.

Another instructive example is given by Wigner. When Heisenberg formulated his quantum mechanics based on matrix calculus, the theory was applicable only to a few idealized problems. Applied to the first real problem, of the hydrogen atom, it also proved successful:

> This was (...) still understandable because Heisenberg's rules of calculation were abstracted from problems which included the old theory of the hydrogen atom. The miracle occurred only when matrix mechanics, or a mathematically equivalent theory, was applied to problems for which Heisenberg's calculating rules were meaningless. Heisenberg's rules presupposed that the classical equations of motion had so-

[41] *Ibid.*, p. 7.

> lutions with certain periodicity properties; and the equations of motion of the two electrons of the helium atom, or of the even greater number of electrons of heavier atoms, simply do not have these properties, so that Heisenberg's rules cannot be applied to these cases. Nevertheless, the calculation of the lowest energy level of helium (...) agrees with the experimental data within the accuracy of the observations, which is one part in ten million. Surely in this case we 'got something out' of the equations that we did not put in.[42]

The second aspect of the mathematicity of the universe may be called *the miracle thesis*. It is possible to imagine worlds which are *mathematical* in a certain sense, yet non-idealizable. Michael Heller considers a hierarchy of such worlds. 'The most non-mathematical' is a world in which no mathematical and logical principles are observed (including any stochastic or probabilistic laws). Next, he suggests to consider a simplified model of the world: let us assume that the world in question may be in one of only two states, represented by '0' and '1'. Now:

> The history of this world is thus a sequence of '0's and '1's. Assume further that the world had a beginning, what may be represented by a dot at the beginning of the sequence. In this way, we get, e.g., a sequence:
>
> .011000101011...
>
> The task of a physicist is to construct a theory which would enable to predict the future states of the world. Such a theory would amount to the 'encapsulation' of the sequence of '0's and '1's in a formula (which is shorter than the sequence it encapsulates). Such a formula may be found only if the sequence of '0's and '1's is algorithmically compressible. But this leads to a problem. Such a sequence may be interpreted as a decimal expansion of a number in [0,1] and – as well known – the set of algorithmically compressible numbers

[42] *Ibid.*, p. 10.

> belonging to [0,1] is of measure 0 (...). Thus (...) there is zero-measure chance that a sequence of '0's and '1's, representing our world, belongs to the set of algorithmically compressible sequences and so the physicist, who investigates such a world, may have no rational expectation to discover the theory she is looking for.[43]

This observation underscores 'the other side' of the mathematicity thesis: not only is the universe mathematical (and hence penetrable by *some* mathematical method), but it is also mathematical in a non--malicious way (and hence penetrable by *our* mathematical methods).

In connection to the problem of the mathematicity of the world, Lakoff and Núñez claim:

> No one observes laws of the universe as such; what are observed empirically are regularities in the universe (...); laws are mathematical statements made up by human beings to attempt to characterize those regularities experienced in the physical universe. (...) What [the physicists] do in formulating 'laws' is fit their human conceptualization of the physical regularities to their prior human conceptualization of some form of mathematics. All the 'fitting' between mathematics and physical regularities of the physical world is done within the minds of physicists who comprehend both. The mathematics is in the mind of the mathematically trained observer, not in the regularities of the physical universe.[44]

This, again, is an example of bad philosophy. Lakoff and Núñez fail to realize the far-reaching consequences of the efficiency thesis. What they leave unaccounted for are, at least, the fact that the mathematical method helped us to conquer the micro-scale phenomena; that equations often 'know more' than their creators; that mathematical models are often the basis for formulating qualitatively *new* predic-

[43] M. Heller, *Czy świat jest matematyczny?*, *op. cit.*, p. 51–52.
[44] G. Lakoff, R. Núñez, *Where...*, *op. cit.*, p. 344.

tions, and so serve as powerful heuristic tools. It seems that behind Lakoff and Núñez's observations there lies a very simplistic or naive view of science: that scientific progress comes from the *observations* of the regularities of real-world phenomena and their generalizations into the mathematically expressible laws of physics. What follows, within Lakoff and Núñez's framework one cannot even formulate the miracle thesis.

Stanislas Deheane offers a more sophisticated argument to explain – within a naturalistic framework provided by the recent findings of neuroscience – the effectiveness of the mathematical method. He notes:

> How can one explain the extraordinary adequacy of the purest products of the human mind to physical reality? In an evolutionary framework, perhaps pure mathematics should be compared to a rough diamond, raw material that has not yet been submitted to the test of selection. Mathematicians generate an enormous amount of pure mathematics. Only a small part of it will ever be useful in physics. There is thus an overproduction of mathematical solutions from which physicists select those that seem best adapted to their discipline — a process not unlike the Darwinian model of random mutations followed by selection. Perhaps this argument makes it seem somewhat less miraculous that, among the wide variety of available models, some wind up fitting the physical world tightly.[45]

The problem with this argument is that it does not touch what is the crux of the efficiency thesis. In the above quoted passage Wigner admits that only a small fraction of our mathematical theories find application in physics; but his astonishment stems from a different source: given that mathematics provides us with "an amazingly accurate description of a large class of phenomena" and that physical equations often 'know more' than their creators, shows "that the mathematical language

[45] S. Deheane, *The Number*..., *op. cit.*, p. 232–233.

has more to commend it than being the only language which we can speak; it shows that it is, in a very real sense, the correct language".

Deheane continues by observing that:

> In the final analysis, the issue of the unreasonable effectiveness of mathematics loses much of its veil of mystery when one keeps in mind that mathematical models rarely agree *exactly* with physical reality. Kepler notwithstanding, planets do not draw ellipses. The earth would perhaps follow an exact elliptic trajectory if it were alone in the solar system, if it was a perfect sphere, if it did not exchange energy with the sun, and so on. In practice, however, all planets follow chaotic trajectories that merely resemble ellipses and are impossible to calculate precisely beyond a limit of several thousand years. All the "laws" of physics that we arrogantly impose on the universe seem condemned to remain partial models, approximate mental representations that we ceaselessly improve.[46]

This is a gross misunderstanding. Of course, mathematical physics takes advantage of idealizations: in constructing mathematical models of reality, the strategy is to disregard a number of aspects of the phenomena under consideration. But this is exactly what has made science possible. Moreover, the fact that idealization is possible says something about the universe: that it has a feature thanks to which it can be efficiently investigated with the use of mathematical method, a method that takes advantage of idealization. This means, however, that idealization is possible *because* the universe is mathematical (in the Heller's sense of the word).

I believe that it is relatively easy to point to one of Deheane's assumptions which prevents him from appreciating that the possibility of idealization constitutes an argument in favour of the mathematicity of the universe. He says that "all the 'laws' of physics that we arrogantly impose on the universe seem condemned to remain par-

[46] *Ibid.*, p. 233.

tial models, approximate mental representations that we ceaselessly improve".[47] Notice the use of the verb 'impose': it suggests that, for Deheane, mathematics is only what our minds construct. With such a definition, the conclusion that the world is not mathematical trivially follows. However, this is not what philosophers have in mind while speaking of mathematical Platonism or the unreasonable effectiveness of mathematics in the natural sciences; they rather claim that our mathematics, i.e. the mathematical theories we have developed, somehow captures or 'resonates' with mathematical reality, be it the Platonic universe of pure abstract objects or some aspect of the physical universe. As Michael Heller puts it:

> It is obviously true that genetically our mathematics comes from the world: we abstract some of its features. However, one needs to carefully distinguish between *our mathematics* and *mathematics as such*. Our mathematics (which I also deem 'mathematics with a small m') has been developed by humans in a long evolutionary process: it is expressed in a symbolic language we invented; its results are collected in our scientific journals, books, or computer memory. But our mathematics is only a reflection of certain relations or structures, which governed the movement of atoms and stars long before biological evolution began. I deem those relations or structures mathematics as such (or 'Mathematics with a capital M'); it is what we think of when we ask, why nature is mathematical. The answer to this question, which posits that the nature is mathematical because mathematics has been abstracted from nature, turns out helpless, or even naïve, when one introduces the distinction between our mathematics and mathematics as such.[48]

Thus, it seems that neither Lakoff and Núñez nor Deheane are able to provide any tenable answer to the efficiency thesis; with regards the

[47] *Ibid.*, p. 239.

[48] M. Heller, *Co to znaczy, że przyroda jest matematyczna?*, [in:] *Matematyczność przyrody*, eds. M. Heller, J. Życiński, Petrus, Kraków 2010, p. 16.

miracle thesis, they do not even formulate it. However, the interesting fact is that the conception of mathematics, which draws on their theories, may shed some light on the efficiency thesis. The argument is quite general. Both our inborn mathematical capacities, as well as our conceptual apparatus have been shaped – in the evolutionary process – by our interactions with the environment. Now, given that our environment is mathematical (in Heller's sense of the word), it helps us to understand why our mathematical concepts are efficient in uncovering the laws of the universe. Of course, such an argument cannot explain fully the efficiency of mathematics in quantum physics, or the fact that physical equations sometimes 'know more' than their creators. However, it may serve to dismiss the idea that "all the 'fitting' between mathematics and physical regularities of the physical world is done within the minds of physicists who comprehend both. The mathematics is in the mind of the mathematically trained observer, not in the regularities of the physical universe". On the contrary: the mind is mathematical because it is a part of the mathematical universe.

The views of Lakoff, Núñez, and Deheane illustrate nicely that the approach of (some) neuroscientists towards philosophical problems is a kind of *scientific foundationalism*: they seem to believe that neuroscience provides us with knowledge freed from philosophical assumptions, when the opposite is true – scientific theories are often intertwined with philosophical contents. For instance, in the above described example, at least the following philosophical doctrines have been embraced by neuroscientists: that an alleged rational intuition is the only mode of grasping abstract objects (this assumption seems to be shared by Lakoff and Núñez, as well as Deheane); that science (including physics) proceeds by induction and abstraction from observations (Lakoff and Núñez); and that the only mathematics one can speak of is the mathematics we constructed (Deheane). I hope to have illustrated that these are not the only philosophical stances one may adopt. This is not to say that there is no controversy here, e.g. that there really exists Mathematics with the capital M or that the assumptions embraced by Lakoff, Núñez and Deheane are untenable;

to the contrary, they are perfectly acceptable *philosophical* claims. The illusion is, however, that these issues have already been settled by neuroscience.

3. Enrichment

I argue that in order to depict the relationship between philosophy and neuroscience one needs to acknowledge that neither discipline is *isolated* from the other, nor provide *foundations* for the other. Argumentation – both in philosophy and in neuroscience – is *non-foundational*.

In his essay "Against Foundationalism" Michael Heller observes that each philosophical argument has two components: the deductive and the hermeneutic:

> I believe that all arguments in philosophy, but also in the sciences, can be arranged in a sequence, such that at its – say – left end there are arguments without the hermeneutic component, while at the right – arguments without the deductive component. (...) Rationalistic arguments are relatively closer to the left-hand side of the sequence; visionary arguments are relatively close to the right-hand side. Crucially, any philosophical argument, which pertains to a non-trivial philosophical claim, is never devoid of the hermeneutic component.[49]

He also adds that:

> in a typical situation there exists a kind of feedback between the vision and the logical argumentation. Even if the chain of arguments is inspired by a vision, rational argumentation may influence it, giving rise to its revisions and, in a critical situation – even to its rejection.[50]

---

[49] M. Heller, *Przeciw fundacjonizmowi*, [in:] M. Heller, *Filozofia i wszechświat*, Universitas, Kraków 2006, p. 93.

[50] *Ibid.*, p. 94.

Such a view of philosophical argumentation leads firmly to the rejection of foundationalism: if argumentation is a constant interplay of the hermeneutic vision and deduction, there exist no indefeasible, 'clear and distinct' premises, or there exists no unshakable foundation of our philosophical constructions. Argumentation in philosophy takes on a different form:

> When one begins to solve a problem, (...), one accepts certain *hypotheses* (...). It is important to note that these are hypotheses, not certainties (...), and maybe even working hypotheses. By using them one arrives at a solution of a problem (…). The results of the analysis may either strengthen one's initial hypotheses, or lead to their modifications. Such a procedure may be repeated multiple times, resulting in the self-adjustment of the system.[51]

Heller's insightful remarks may be summarized – and given more precise form – in the following way. Any philosophical argumentation must meet four conditions:

(a) the revisability condition: at least some of the premises of any philosophical argumentation are hypotheses – they can be rejected or modified;
(b) the feedback condition: the modification or rejection of premises (hypotheses) must be based on the evaluation of their logical consequences;
(c) the background stability condition: the argumentation background (some previously accepted theories other than the evaluated hypotheses) is relatively stable in relation to the hypotheses; it should be easier to modify or reject the hypotheses than the background;

[51] M. Heller, *Nauki przyrodnicze a filozofia przyrody*, [in:] M. Heller, *Filozofia i wszechświat,* Universitas, Kraków 2006, p. 32.

(d) the disputability condition: any philosophical argumentation is open to formulating competing, even contradictory, hypotheses.

Heller rightly observes that arguments that meet the above stated conditions cannot be accounted for within classical logic. He urges us therefore to look for a 'non-linear logic', or such a logic that would encapsulate the structure of non-foundational thinking.[52] Although I cannot offer such a full-blooded logic here, I would like to suggest that non-foundational arguments can be explicated with the use of some non-classical but well-known formal tools and, in particular, the formal theory of belief revision and the formal theory of coherence.

The idea is simple: with a given language *L* and the background knowledge *K* one puts forward certain hypotheses *H1, H2, H3, ...,* each aiming at solving a problem at hand. We shall say – simplifying considerably – that a problem is defined by a pair of contradictory sentences {*p*, ~*p*}, and that to solve a problem means to determine which of the sentences, *p* or ~*p*, is true. Thus, a hypothesis *H* solves a problem when it (together with some other previously accepted sentences) implies *p* or ~p. Importantly, any newly introduced hypothesis *H* together with the background knowledge *K* may yield contradiction. In such cases, one needs to revise or reject some parts of the background knowledge, and this procedure is well modeled in formal theories of belief revision.[53] In other words, the set *K*H1*, i.e., *K* revised by *H1*, may not include every sentence, which was originally in *K* (I simplify here, disregarding the fact that there usually are many ways of revising *K* by *H1*, and so the set *K*H1* is in fact chosen from among the possible ways of modifying *K* in order to accommodate

[52] The classical relation of logical consequence is a non-linear function. In addition, there exist formal systems called nonlinear logics. However, Heller speaks of something different – a logic of epistemological non-foundationalism – and hence I used the term 'non-linear logic' in quotation marks.

[53] Cf. P. Gardensors, H. Rott, *Belief Revision*, [in:] *Handbook of Logic in Artificial Intelligence and Logic Programming, vol. IV: Epistemic and Temporal Logic*, eds. D.M. Gabbay, Ch. Hogger, J.A. Robinson, Oxford University Press, Oxford 1995, pp. 35–132.

H1). To put it succinctly: revisions such as *K*H1, K*H2, K*H3* often result in the modifications to the background knowledge.

Whether such modifications are acceptable depends on whether an introduced hypothesis (*H1*, *H2, H3*) indeed solves a problem that has previously remained unsolved. However, this is not the only criterion for assessing the quality of a hypothesis. The other such criterion is coherence: we shall say that the better the hypothesis (solving some problem) is, the more coherence it generates in our system of beliefs. Coherence is determined by taking into account: (a) the number of nontrivial inferential connections in our belief set (so in *K*H1, K*H2, K*H3* respectively); and (b) the degree of its unification.[54] There exist nontrivial inferential connections between sentences belonging to a given set if they can serve together as premises in logically valid schemes of inference. In turn, a given set of sentences is unified if it cannot be divided into two subsets without a substantial loss of information.

Thus, the question is, which from among the considered hypotheses *H1, H2,* and *H3* (of which all solve the problem at hand), should be given priority? The answer lies in the interplay between two factors: the extent of modifications a hypothesis causes within our background knowledge (the less changes the better), and the degree of coherence it brings about in our belief set (the higher degree the better). There is no simple formula to settle this interplay, it is rather a matter of decision on case by case basis. However, it is reasonable to assume that if two hypotheses, *H1* and *H2*, bring about a similar level of coherence, and when *H1* causes substantial modifications in the background knowledge, while *H2* changes it only slightly, it is *H2* that should be preferred. Similarly, when both hypotheses produce similar modifications in the background knowledge, but one of them brings about more coherence, it should be preferred. It must also be added that there may be situations in which *all* of the considered hy-

[54] L. Bonjour, *The Structure of Empirical Knowledge*, Harvard University Press, Cambridge, Mass. 1985.

potheses cause so substantial changes to the background knowledge that they cannot be accepted, even if they solve the problem at hand and bring about much coherence.

The situation depicted above, i.e. one which takes into account only the background knowledge and the hypotheses, is a substantial simplification. However, it may easily be extended to give a more fine-grained description of non-foundational argumentation. For instance, one can utilize the concept of presuppositions, which enables to capture two important aspects of non-foundational thinking. Firstly, one can speak of the presuppositions *P* of the background knowledge *K*; in particular, the set *P* may contain the so-called existential and lexical presuppositions. Existential presuppositions posit the existence of a certain entity or a situation (e.g., when I say that "John has a new car" it presupposes that John exists); lexical presuppositions, on the other hand, are sentences which must be true in order for some concepts to be applicable (the lexical presuppositions of the sentence "John is not a bachelor" include "John is a male"). The introduction of the set of presuppositions *P* enables one to describe a situation in which a hypothesis leads to the modification not only of some fragment of our background knowledge, but also of our existential commitments and our conceptual scheme (when it causes the rejection of an existential or a lexical presupposition, respectively).

Secondly, the utilization of the concept of presupposition enables one to account for a situation in which one determines that a given problem is ill-stated. This requires a modification in the way we understand the process of solving problems. We shall say that a hypothesis *H* solves a given problem defined by the set $\{p, \sim p\}$ if *H* (possibly together with some other sentences belonging to the background knowledge) deductively implies *p* or $\sim p$, or it deductively implies $\sim s$, where *s* is a presupposition of *p*. In the latter case – where a presupposition of *p* turns out false – one can say that the solution to the problem defined by the pair $\{p, \sim p\}$ is that the problem is ill-stated, i.e. neither *p* nor $\sim p$ can be ascribed truth-values.

The introduction of presuppositions into our formal account of non-foundational argumentation requires two additional comments. The first is that while our background knowledge should be more stable (i.e., more immune to revisions) than our hypotheses, our presuppositions should be more stable than our background knowledge. Thus, when one chooses from among a number of hypotheses of which all solve the problem at hand and bring about much coherence into one's belief set, the hypothesis should be preferred which causes fewer modifications within one's system of presuppositions. Still, it must be stressed that taking advantage of the mechanism of presuppositions requires changes in the logic underlying non-foundational reasoning.[55]

The above described procedure meets all the conditions of non-foundational argumentation. Firstly, neither the hypotheses one considers, nor one's background knowledge, are immune to revisions, and so the revision condition is fulfilled. Secondly, the quality of hypotheses hangs together with the changes they bring about in our belief system, and they are modified or rejected if the changes are unacceptable (so, the feedback condition is met). Thirdly, the background stability condition is fulfilled since although background knowledge is not immune to revisions, from among the hypotheses that solve the problem and bring about a similar level of coherence, the one that should be preferred is the one that saves most of the original background knowledge. Moreover, in cases when all the hypotheses cause substantial modifications of the background knowledge, they may all be rejected. Fourthly, as the above described formal framework enables one to work simultaneously with several hypotheses, the disputability condition is met (it must be stressed, however, that this requires a special underlying logic, e.g., the so-called defeasible logic).

In the passage quoted above, Michael Heller suggests that non-foundational thinking is typical not only in philosophy, but also in

[55] Cf. B. van Fraassen, *Presupposition, Implication and Self-Reference*, "The Journal of Philosophy" 1968, vol. 65(5), pp. 136–152.

science. This becomes clearly visible when one considers the structure of argumentation in neuroscience. Let us begin with an idealization: although it is commonly accepted in the philosophy of science that there exist no theory-free observations and experiments, and that our theories play important heuristic and interpretation roles in our observational and experimental activities, let us assume that there are 'pure' neuroscientific facts (results of observations and outcomes of experiments). What does a neuroscientific explanation of such facts consist of? I posit that there are three different criteria at work here: empirical adequacy, convergence and coherence. An empirically adequate theory must connect facts in such a way that it may serve as a means of prediction (even if not an infallible one). For instance, a theory that posits the existence of an inborn Object Tracking System, which is capable of discriminating up to 4 objects, would be empirically inadequate if it turned out that infants are capable of tracking 10 or 15 object at once.

Still, there may exist various competing theories explaining the same set of facts. For instance, as I indicated above, there is a controversy regarding how children move from using numbers 1–4, which seems to be an innate skill, to mastering arithmetic. One proposal was put forward by Piazza.[56] She observes that the Approximate Number System (ANS) may be used to represent not only large numbers, but also small ones. Moreover, ANS quite quickly becomes very precise as regards small numerosities. Given the progression in the sensitivity of ANS, in order to distinguish between 2, 3, and larger numbers a ratio of 3:4 is needed. This happens at around three years of age, and coincides with the period when children become 'three-knowers'. In other words, Piazza believes that no interplay between OTS and ANS is needed to 'break the number four barrier' – the increasing precision of the ANS system is sufficient to account for this ability.

Another hypothesis which addresses this problem is defended by Spelke. She observes that "children appear to overcome the limits

[56] M. Piazza, *Neurocognitive...*, *op. cit.*, p. 275–276.

of the core number system when they begin to use number words in natural language expressions and counting".[57] Children learn the first ten counting words by the age of 2, but initially use them without the intended meaning. At three they know that 'one' means one; at four they associate '2', '3' and '4' with the corresponding numerosities. Then, there is a kind of 'jump' – children learn quite quickly the next numbers. This, according to Spelke, requires two things: (a) to understand that every word in the counting list designates a set of individuals with a unique cardinal value; and (b) to grasp the idea that each cardinal value can be constructed through progressive addition of 1.[58] How this is possible? "For most children, the language of number words and verbal counting appears to provide the critical system of symbols for combining the two core systems (i.e., ANS and OTS), and some evidence suggests that language may be necessary for this construction".[59]

Thus, we have two competing explanations of the same set of facts: that human innate skills cannot account for simple arithmetic, and that something in individual development must facilitate – or even enable – 'breaking the number 4 barrier'. Piazza believes that the increasing sensitivity of ANS is sufficient to explain how it happens, while Spelke claims that it is the development of the language skills that plays the pivotal role here. How to decide which of those hypotheses is acceptable? One of the possibilities is to use the criterion of convergence. Let us recall some other facts, referred to above. First, both children and adults in remote cultures, whose languages have no words for numbers, when dealing with numbers larger than three only recognize their equivalence approximately. Second, deaf persons living in numerate cultures but not exposed to the deaf community use a gestural system called homesign; they use fingers to communicate numbers, but only with approximate accuracy. Third,

[57] E.S. Spelke, *op. cit.*, p. 304.
[58] *Ibid.*, p. 305.
[59] *Ibid.*, p. 305.

educated adults who suffer language impairments have problems with exact, but not approximate numerical reasoning. Fourth, when doing exact (but not approximate!) tasks adults spend more time with numbers that are difficult to pronounce, even if they are presented in Arabic notation. Fifth, bilingual adults who are taught some new mathematical facts in one of their languages have difficulties in the smooth production of exact number facts in the other language.[60] All these facts support Spelke's hypothesis – but not Piazza's – because it is Spelke's claim that language is essential to acquiring arithmetic skills that is empirically adequate for a larger set of facts. In other words, Spelke's hypothesis *converges* on more experimental and observational data than Piazza's.

Another criterion that may be used to pick from among competing – and empirically adequate – hypotheses is that of coherence. Spelke's claim that language is essential in the development of arithmetic skills seems highly coherent with Lakoff's theory of embodied mind (as presented above), while Piazza's hypothesis is not. This may be seen as an argument from coherence in favour of Spelke's hypothesis. At the same time, there may be other theories – e.g., some incarnations of the modular mind paradigm – which would favor Piazza's stance. The point is that the criterion of coherence constitutes an important justification standard in neuroscientific discourse.

Thus, even in the idealized picture of neuroscientific practice we assumed, one that posits the existence of pure, theory-free facts, there are *competing explanations* of the same set of facts, and the criteria for choosing from among them include convergence and coherence. Of course, the situation becomes even more complicated when we drop our idealizing assumption and admit that our theories – and, in particular, entire paradigms, such as embodiment or the modular paradigm – provide both heuristic and interpretation frameworks for neuroscientific practice. But the conclusion remains the same: neuroscientific thinking *is*, at its core, *non-foundational*. Neuroscientific

---

[60] *Ibid.*, p. 307.

hypotheses – as well as background knowledge – are revisable, and the revisions are caused not only by empirical inadequacy of our theories, but also by the consequences we draw from our new hypotheses. Background knowledge in neuroscience is usually quite stable (as illustrated by the persistence of entire paradigms, such as the embodied or modular one). Finally, neuroscientific argumentation fulfils the disputability condition: one usually formulates and chooses from among a number of hypotheses explaining the given phenomenon.

The fact that both philosophy and neuroscience take advantage of non-foundational reasoning has far-reaching consequences for any description of the interactions between them. Since neither discipline provides foundations for the other, the interactions in question are never uni-directional. Generally speaking, such mutual interactions are possible because some neuroscientific theories may constitute (a part of) the background knowledge or presuppositions of philosophical conceptions, and some philosophical theses may be elements of the background knowledge, or presuppositions of neuroscientific theories. It must be stressed, however, that it is only so if both philosophical conceptions under consideration and the given neuroscientific theories take advantage of the same language, or at least the languages they use are (partially) mutually translatable.

In a previously published paper I suggested that the interactions between philosophy and neuroscience take place at four different levels: the conceptual level, the presuppositional level, the problem level and the functional level.[61] At the conceptual level, the conceptual schemes of both philosophy and neuroscience may overlap, and concepts from one discipline may migrate to the other. This can be accommodated within the framework developed in this essay. Moreover, the framework enables also *conceptual change* brought about by the hypotheses formulated in one discipline leading to the rejection of the lexical presuppositions of the other. At the presuppositional level, the theories developed in one discipline may lead to the claim

[61] Cf. B. Brożek, *Philosophy in Neuroscience*, *op. cit.*

that some sentences within the other discipline transpire to be meaningless (e.g., when a neuroscientific theory revises some existential presuppositions of a philosophical conception). Given the above described framework, philosophy and neuroscience may pertain to the same *problems*, or one may generate problems taken up by the other. It is also possible that some existing problems may become ill-stated, when one of the disciplines (e.g., neuroscience) provides us with the grounds to reject a presupposition of some (e.g. philosophical) problem. Finally, at the functional level, both disciplines may provide heuristic tools as well as justification criteria for the other – ultimately, it is not *a priori* decidable what constitutes one's background knowledge.

Of course, it is not to say that there is no difference between neuroscience and philosophy: they both have their own methods, justification criteria and typical problems, which are built-in into neuroscientific or philosophical practice. The point is not that any belief is equally good (there may be better or worse arguments both in neuroscience and philosophy), but that there is no *critical* structural difference between the two types of argumentation. In both cases one has to do with non-foundational reasoning, and this is exactly what enables mutual enrichment between philosophy and neuroscience.

Robert Poczobut
University of Białystok

Mind, Emergence, and the Limits of Neuroscientific Explanations[1]

> Naturalizing the conscious mind and planting it firmly in the brain does not diminish the role of culture in the construction of human beings, does not reduce human dignity, and does not mark the end of mystery and puzzlement.
>
> *Antonio Damasio*

Introduction

Over the past several decades we have been witnessing an intellectual ferment that might be called, without exaggeration, a *neuroscientific revolution*. The findings obtained by using new experimental techniques, which enable researchers to study a live brain while it is engaged in mental/cognitive activity, are being applied in nearly all areas of life. Both the human body and the mind depend on the efficient functioning of the brain for their activity. The brain controls all bodily organs and their physiological processes, and it is responsible for the realization of mental/cognitive states and capacities, including consciousness, intelligence, memory, decision-making, and artistic expression. The implications of the neuroscientific revolution

---

[1] This paper was written within the research grant "The Limits of Scientific Explanation" sponsored by the John Templeton Foundation.

that we are experiencing are not merely of practical, technological, or medical value; their significance is above all epistemic. A spectacular increase in knowledge about the brain offers us new insights into ourselves, leading to the discovery of new levels of organization in man, and allowing us to redefine a number of traditional philosophical notions and categories (the person, the self, consciousness, free will, agency). What I shall attempt in this paper is to define the tasks and goals of philosophy as it is practiced in the context of neuroscience, and to call attention to the limits of accounting for mental/cognitive processes solely in terms of neural mechanisms.

1. Philosophy in the Context of Neuroscience

1.1. Contemporary neuroscience is a multidisciplinary research program which concerns itself with analyzing various aspects of the operation of the nervous system (its structure and functions, its evolutionary and developmental history, its physiology, anatomy, and pathology). Like other scientific disciplines, neuroscience – the disciplines and theories it comprises – makes particular ontological commitments (about the mind-brain-environment relations) and methodological assumptions (about the nature and scope of research methods). The immediate aim of neuroscience is to explain the behaviours of man and other animals in terms of the activity of their brain and nervous system or, in other words, to answer the question: How do millions of nerve cells generate an organism's complex adaptive and cognitive behaviours, and how, in turn, does the environment contribute to reshaping them? A more distant objective is to account for the neurobiological and neurocomputational mechanisms of consciousness, as well as of other mental/cognitive functions, including perception, learning, memory, and abstract thought.[2]

[2] E.R. Kandel, J.H. Shwartz, T.M. Jessel, *Principles of Neural Science*, McGraw-Hill, New York 2000.

The number of disciplines whose names begin with the prefix "neuro" is impressive. Neuroscience encompasses: neuroanatomy, neurophysiology, neurophysics, neurochemistry, neuropsychiatry, neuropsychology, cognitive neuroscience, social neuroscience, computational neuroscience and neuroinformatics, the theory of neural networks, evolutionary neuroscience, developmental neuroscience, neuroendocrinology, neuroengineering, neurocybernetics, neuroeconomics and neuromarketing, neurolinguistics, neuropsychopharmacology, neuroimaging, neurogenetics, cellular and molecular neuroscience, behavioural neuroscience, neuropathology, neuroimmunology, neurophilosophy (neurophenomenology, neuroethics, neuroaesthetics), and neurotheology. The fundamental neurosciences (neuroanatomy and neurophysiology) pose questions of the type: How does a particular neurobiological system realize process (function) X? The answer usually takes the form of a mechanistic explanation, which proceeds by identifying the relevant micromechanisms underlying the process under investigation. By contrast, the principal questions of evolutionary neuroscience begin with the word "why" – whereas the answers take the form of historical explanations in terms of natural selection. In short, while the fundamental neurosciences seek to describe immediate, or proximate, causes (mechanisms), while evolutionary neuroscience focuses on ultimate causes. In order to provide a comprehensive account of a neurobiological phenomenon, one must identify both the immediate and the ultimate causal connections involved.[3]

The neuroscientific disciplines enjoying the most rapid development these days include cognitive neuroscience. It aims to explain how cognitive processes arise and develop (the diachronic aspect) and how they are realized by brain activity (the synchronic aspect). Contrary to an older approach dominating cognitive science, researchers now do not draw a sharp distinction between the theory of information processing and brain theory, which treats the brain as a neurobiological

---

[3] A. Revonsuo, *Inner Presence: Consciousness as a Biological Phenomenon*, MIT Press, Cambridge, Mass. 2006, ch. 1; E. Mayr, *This is Biology: The Science of the Living World*, Harvard University Press, Cambridge, Mass. 1998, ch. 4.

mechanism. It is supposed that one day cognitive neuroscience will be able to discover algorithms leading from structurally simple neural elements to global neurophysiological activity resulting in perception, memory, and consciousness.[4]

Cognitive neuroscience treats the level of mental/cognitive phenomena as a higher level of brain organization; it is to be explained in terms of lower-level neural and neurocomputational mechanisms. Although mental/cognitive processes do not constitute an explanatorily autonomous domain (they do not account for themselves), they are not subject to displacement or elimination by processes from the lower level. This is typical of the majority of the special sciences investigating higher-level processes. Such sciences are not fully autonomous, nonetheless the processes they study are as real as the fundamental physical forces. Such is the status of cognitive psychology, but also of sciences like cytology, which is not subject to displacement or elimination by the lower-level disciplines, even though it is not completely independent of them.

1.2. It is important in the context of our discussion that neuroscience is itself part of a considerably broader multidisciplinary research program known as cognitive science. The core of contemporary cognitive science is made up of such disciplines as: computer science and information theory, cognitive neuroscience, cognitive psychology, cognitive linguistics, artificial intelligence and cognitive robotics, as well as philosophy of mind and disciplines studying the influence of socio-cultural phenomena on cognitive processes. After a period dominated by computer science and neuroscience, cognitive science has taken an interest in factors belonging to the exostructure of cognitive systems (their surroundings, context, environment, developmental and evolutionary history, socio-cultural influences). This development was brought about by a realization that computational, neural, and molecular mechanisms are not the only mechanisms involved in cog-

[4] M. Gazzaniga, *The Cognitive Neuroscience*, MIT Press, Cambridge, Mass. 1995.

nitive processes. Given that humans are social animals, most mental processes involve complex interactions with other members of mankind. Learning, decision-making, or problem-solving do not reduce to processes at the molecular level, because these processes presuppose a contribution from higher-level mechanisms, namely psychological and social ones. The aspects (dimensions, levels) of mental processes targeted by explanations invoking these higher-level mechanisms differ from those that hold an interest for a molecular biologist. The psychological and sociological accounts of the mind should be treated as complementary to, rather than incompatible with, explanations that appeal to chemical processes occurring in the cells of our bodies. Mental/cognitive processes do not constitute a homogeneous domain, but form an extremely diverse complex of problems.[5]

According to Philip Johnson-Laird, cognitive science came into being once researchers realized that no monodisciplinary approach could ever account for all aspects of human mind. This realization of the multidimensional nature of mental/cognitive processes had come gradually. Initially, workers in a number of disciplines, using different research methods, took up problems concerning various aspects of the functioning of the mind. Computer scientists constructed computational models of mental activity; linguists studied the mental mechanisms of linguistic competence; ethologists looked for innate patterns of social behaviours; neurophysiologists analyzed the relations between nerve-cell functions and complex cognitive capacities; neuropsychologists examined damaged brains to identify the neural correlates of mental activities; cognitive psychologists focused on perception, memory, thinking, and learning; evolutionists attempted to reconstruct the phylogeny of natural mental/cognitive systems; anthropologists investigated the conceptual structures underlying complex cultural behaviours; philosophers of mind (mainly analytic

[5] W. Bechtel, A. Abrahamsen, G. Graham, *The Life of Cognitive Science*, [in:] *A Companion to Cognitive Science*, eds. W. Bechtel, G. Graham, Blackwell, Oxford 1998, pp. 1–104.

philosophers and phenomenologists) conducted detailed analyses of consciousness, intentionality, mental contents, and propositional attitudes.[6] It was only after the convergence of these disciplines and the interrelations between their domains had been fully appreciated that cognitive science could emerge – as a multidisciplinary scientific endeavour concerned with the origin, structure, principles of organization and mechanisms of operation of mental/cognitive systems: both animal, human, and artificial.

One methodological peculiarity of research into mind and cognition, which is without precedence in the history of science, is that, as it turns out, unravelling the secrets of the mind requires tapping into the theoretical resources of logic, mathematics, the natural sciences, engineering, medicine, as well as the humanities and philosophy. A considerable majority of researchers believe that mental/cognitive processes are irreducibly multidimensional in character, extending well beyond the boundaries of the research areas and explanatory resources of isolated disciplines.

Despite this programmatic diversity, theoretical cognitive science is developing categories that afford a relative unification of its subdisciplines. One such comprehensive unification project, which had originally served as a foundation for cognitive science, attempted to model, in a uniform fashion, all mental/cognitive processes in terms of receiving and processing information. In the beginning, it was assumed that information in a cognitive system is coded as symbolic representations which undergo transformations according to a defined set of rules (usually, expressible as algorithms). Interestingly, the details of how these rules and representations are implemented (i.e., the physical makeup of their material vehicles) were considered irrelevant. Later on, however, the computational perspective became assimilated into neuroscientific research, and, nowadays, computational models of brain functions are being successfully constructed

[6] Ph. Johnson-Laird, *The Computer and the Mind: An Introduction to Cognitive Science*, Harvard University Press, Cambridge, Mass. 1988, p. 7.

within computational cognitive neuroscience.[7] What we are seeing, then, is a convergence of brain theory and certain results from computer science, which is a necessary condition for developing specific interdisciplinary programs in neuroinformatics and neurocybernetics.

1.3. The slogan *philosophy in the context of neuroscience* refers to two closely related research projects. The first, known as *neurophilosophy,* is substantive in character, while the second, called *the philosophy of neuroscience,* is metascientific (methodological).

Neurophilosophy aims to apply the latest neuroscientific findings to an analysis of classical philosophical topics such as perception, consciousness and the self, personal identity, memory, free will, the moral sense, religious and aesthetic experience, or knowledge of the alter ego (research on mirror neurons). Neurophilosophy has become a bridge discipline between neuroscience and philosophy of mind.[8] It divides into: the neurophilosophy of consciousness,[9] neuroethics,[10] neuroaesthetics,[11] and social neurophilosophy.[12]

A great majority of neurophilosophers assume that neuroscience furnishes information which may contribute to a deeper understanding or even the resolution of many of the traditional puzzles of the

---

[7] W. Duch, *Brain-inspired Conscious Computing Architecture*, "Journal of Mind and Behaviour" 2005, vol. 26 (1–2), pp. 1–22.

[8] H. Walter, *Neurophilosophy of Free Will: From Libertarian Illusions to a Concept of Natural Autonomy*, MIT Press, Cambridge, Mass 2001.

[9] A. Revonsuo, *Inner Presence: Consciousness as a Biological Phenomenon*, MIT Press, Cambridge, Mass. 2006.

[10] M.J. Farah, ed., *Neuroethics: An Introduction with Readings*, MIT Press, Cambridge, Mass. 2010; P.S. Churchland, *Braintrust: What Neuroscience Tells us about Morality*, Princeton University Press, Princeton 2011.

[11] S. Zeki, *Splendors and Miseries of the Brain, Love, Creativity and the Quest for Human Happiness*, Wiley-Blackwell, London 2009; M. Skov, O. Vartanian, *Neuroaesthetics*, Baywood Publishing Company, Amityville 2009.

[12] L. Brothers, *The Social Brain: A Project for Integrating Primate Behaviour and Neurophysiology in a New Domain*, [in:] *Foundations in Social Neuroscience*, ed. J. Cacioppo et al., MIT Press, Cambridge, Mass. 2002, pp. 366–387; G. Rizzolatti, C. Sinigalia, *Mirrors in the Brain: How Our Minds Share Actions and Emotions*, Oxford University Press, Oxford 2008; M. Iacoboni, *Mirroring People: The New Science of How We Connect with Others*, Farrar, Straus and Giroux, New York 2008.

philosophy of mind. This forces philosophers to supplement their traditional methods of philosophical analysis with the research tools of modern neuroscience. According to a central tenet of neurophilosophy, an investigation of mental functions that abstracts away from their neurobiological basis yields inadequate models of the mind. It is therefore necessary to study the whole mind-brain; i.e., to analyze mental functions in conjunction with their corresponding brain activities. On the most ambitious conception of neurophilosophy, the philosopher and the neuroscientist become collaborators – who formulate research problems, form explanatory hypotheses, and conduct experiments together.[13] On another view, the philosopher provides a panoramic perspective on neuroscientific findings, and constructs a scientific metaphysics of mind by way of synthesizing the achievements of the various disciplines making up neuroscience. The goal of such synthesis is to replace the naïve (everyday and philosophical) model of the mind by a model built in the context of (neuro)science.[14]

Philosophy of neuroscience, on the other hand, is a specialized branch of the philosophy of science. Given that, for any science X, there exists a philosophy of X, it is hardly surprising that, alongside such disciplines as philosophy of mathematics, philosophy of physics, and philosophy of biology, we also encounter philosophy of neuroscience and philosophy of cognitive science.[15] The research problems of philosophy of neuroscience are mainly metascientific (methodological) in character – they focus on the explanation procedures and experimental methods employed in neuroscience.[16] Special problems

---

[13] P.S. Churchland, *Neurophilosophy: Toward a Unified Science of the Mind-Brain*, MIT Press, Cambridge, Mass. 1986; P.S. Churchland, *Neurophilosophy: The Early Years and New Directions*, "Functional Neurology" 2007, vol. 22 (4), pp. 185–195; P.M. Churchland, *Neurophilosophy at Work*, Cambridge University Press, Cambridge, Mass. 2007.

[14] H. Walter, *op. cit.*; P.S. Churchland, *The Impact of Neuroscience on Philosophy*, "Neuron" 2008, vol. 60, pp. 409–411.

[15] M. Bennett, P. Hacker, *Philosophical Foundations of Neuroscience*, Blackwell Publishing, Oxford 2003; J. Bickle, *Philosophy and Neuroscience: A Ruthlessly Reductive Account*, Kluwer, Dordrecht 2003.

[16] P. Machamer, R. Grush, P. McLaughlin, eds., *Theory and Method in the Neurosciences*, University of Pittsburgh Press, Pittsburgh 2001.

of the former type include questions about what relations obtain between the different levels of description, analysis, and explanation, on the one hand, and the levels of brain organization (from the molecular level to the level of complex functional systems), on the other. Issues arising in this context concern the nature and limits of accounting for processes that occur at the higher levels of brain organization in terms of lower-level theories and concepts, which is a special case of the classical problem of reduction and emergence.[17] Problems of the latter type concern the epistemic evaluation of findings obtained through new experimental methods, which enable us to study the brain in action, while it performs mental/cognitive tasks.[18]

Some problems of neurophilosophy and philosophy of neuroscience are analyzed both at the metascientific (methodological) and the substantive level. The issue of interlevel relations provides a case in point, since one can interpret these relations either as holding between various levels of description, analysis, and explanation (the metascientific aspect), or between states and processes occurring at different levels of organization in the brain, or in another cognitive system (the substantive aspect). In the next sections, we shall take a closer look at how discussions of interlevel relations appeal to different notions of reduction and psychophysical emergence.

2. Psychophysical Emergence

2.1. The basic notion of emergence applies to phenomena that arise from, or depend upon some more fundamental phenomena and that, at the same time, retain a degree of autonomy vis à vis their base. Ontologically speaking, the term "emergence" refers to a synchronic

[17] C.F. Craver, *Explaining the Brain: Mechanisms and the Mosaic Unity of Neuroscience*, Clarendon Press, Oxford 2007; W. Bechtel, *Mental Mechanisms: Philosophical Perspectives on Cognitive Neuroscience*, Routledge, London 2008.

[18] R. Cabeza, A. Kingstone, eds., *Handbook of Functional Neuroimaging of Cognition*, MIT Press, Cambridge, Mass. 2006.

or diachronic relation holding between levels of organization in the world – i.e., between the properties, processes, patterns, and regularities characteristic of particular levels. The synchronic and diachronic aspects of emergence constitute two complementary dimensions of a single phenomenon.[19] One indicator of diachronic emergence is the appearance in an evolutionary process of qualitatively novel structures, properties, functions, regularities, information, etc. Evolutionary processes, occurring on a cosmic scale, generate systems of increasing complexity: from elementary particles and atoms, through macromolecules and multicellular organisms, to complex mental/cognitive systems, cultural patterns, and intentional products of the human mind.[20] A minimum condition of synchronic emergence is that a system at a given instant exhibit systemic processes and properties which are not instantiated at the level of its microstructure. By going beyond subsequent thresholds of organization, natural systems acquire functions and properties which are absent from the level of their constituents.[21]

Much inspiration for philosophical reflection on emergence has come from chaos theory and nonlinear thermodynamics. These theories, called the physics of creative processes, have transformed our understanding of the universe so radically that the opposition between the mental and the physical is no longer taken to capture a fundamental division in nature.[22] The following questions used to appear paradoxical: How can a cluster of inanimate atoms be alive? How can a system that consists of unconscious elements be conscious? Or, to put the problem in more general terms, how can a system (structure) possess properties and perform functions that its constituents do not

---

[19] M. Bedau, P. Humphreys, eds., *Emergence: Contemporary Readings in Philosophy and Science*, MIT Press, Cambridge, Mass. 2008.

[20] H. Morowitz, *The Emergence of Everything: How the World Became Complex*, Oxford University Press, Oxford 2002; M. Silberstein, J. Geever, *The Search for Ontological Emergence,* "The Philosophical Quarterly" 1999, vol. 49, pp. 182–200.

[21] M. Bunge, *Emergence and Convergence. Qualitative Novelty and the Unity of Knowledge*, University of Toronto Press, Toronto 2003.

[22] A. Scott, *Stairway to the Mind*, Springer-Verlag, New York 1995.

possess or perform, respectively? Similar doubts were raised about evolution and development (diachronic emergence): How can inanimate matter give rise to life? Because the idea of an evolutionary transition from state not-A to state A had an air of paradox to it, a priori arguments were common in the history of philosophy to the effect that emergent qualitative change is logically impossible, or that its acceptance would force us to acknowledge such entities as realized contradictions (paraconsistent metaphysics of change).

In the absence of conceptual tools capable of describing this sort of phenomena, philosophers resorted to various mythological solutions: positing vital forces, or mental substances that were supposed to endow a purely deterministic physical machine with life, consciousness, and creativity. However, in light of such successful theories as nonlinear thermodynamics, the theory of self-organizing dynamical systems, information theory, and complex systems theory, it has become clear that systems can possess properties and functions which differ dramatically from the properties and functions of their constituent parts.[23] The apparent dichotomies based on inadequate concepts and theories have thus been overcome, and the ground has been prepared for a new construal of the mind-body problem. As a result, our conception of physical reality has changed, along with our understanding of the relationship between mental phenomena and their neurobiological bases.

2.2. The notion (theory) of emergence captures certain aspects of the relations holding between levels of organization in nature. Some features of these interlevel relations may vary depending on the character of the system under consideration, which may be physical, biological, mental/cognitive, social, or cultural. But the existence of differences does not preclude the existence of similarities. Emergence occurs in multilevel hierarchical systems where each level of organization possesses its own properties and regularities, which are described in the

[23] K. Mainzer, *Thinking in Complexity*, Springer, Berlin-Heidelberg 2003.

theoretical language of a particular order. Each level in a hierarchical system exhibits a unique set of causal connections that cannot be described – without loss of information – in the language of lower-level theories. Emergence arises from a multitude of causal-informational interactions and feedback mechanisms, including bottom-up, top-down, and same-level interactions.[24]

Emergence may consist in the appearance of a new species, individual, or function (activity, behaviour) realized by an individual. All kinds of emergence appear as a result of the adaptive selection that occurs in interaction with a physical, biological, or social environment which determines the system's boundary conditions. The emergence of complex systems from simpler physical structures is possible due to nonlinear interactions between their constituent parts, on the one hand, and between the system and its environment, on the other (this is the nonlinear, contextual dimension of emergence). Live systems are structured so as to be able to control their behaviour via feedback mechanisms, and learn by ways of (a) receiving, storing, and processing information, (b) recognizing patterns, and (c) building internal models of the environment. Because emergent phenomena characteristic of the level of the mental (consciousness, the ability to abstract, as well as having beliefs, concepts, plans, intentions, goals, etc.) are causally efficacious in a world of fields, particles, and physical forces, we have reason to treat them as equally real.[25]

Although we still lack a complete understanding of the principles of self-organization, no one these days attempts to explain the structure of an atom, snowflake, DNA, or consciousness by invoking entelechies, *elan vital*, or *res cogitans*. All the laws (programs) necessary to generate the wealth of structures present in the universe originate in nature. Simple, regular structures (like crystals) are generated by relatively simple laws (programs), whereas structures characterized

[24] N.A. Baas, C. Emmeche, *On Emergence and Explanation*, "Intellectica" 1997, vol. 2, pp. 67–83.

[25] G.F.R. Ellis, *On the Nature of Emergent Reality*, [in:] *The Re-Emergence of Emergence*, eds. P. Clayton, P. Davies, Oxford University Press, Oxford 2006, pp. 79–110.

by *logical depth* (a term introduced by Charles Bennett in the context of algorithmic information theory) can only develop as a result of correspondingly complex processes. One reason biological and mental/cognitive systems have logical depth is that they rely for their development on a long and complex evolutionary chain.[26]

The discovery of genuinely creative processes in nature permits the partial integration of the worldview based on the natural sciences and that developed in the humanities. The creativity of the universe became considerably amplified when mankind entered the stage. Human minds generate a world of objective meanings, values, norms, abstract entities, and theories. This, however, was just a continuation of an existing line of development. Conscious minds constitute an integral part of the universe, and could not have emerged, or acquired all the characteristics they possess, had it not been for the right environmental and structural-nomological conditions. According to Philip Clayton:

> The distinctiveness of the emergentist thesis lies in its claim that the natural world exhibits a variety of levels at which distinct types of laws and causes can be recognized. [...] Understanding the relationship between mind and brain – between consciousness and its neural correlates – requires understanding the multi-leveled structure of the natural world. On this view, the appearance of mental causes is, in one sense, just another case of emergence – just another case in which a complicated natural system gives rise to unexpected causal patterns and properties.[27]

2.3. Advocates of psychophysical emergentism answer the following questions in the affirmative: (a) Is the mind an object of scientific

---

[26] Ch. Bennett, *Logical Depth and Physical Complexity*, [in:] *The Universal Turing Machine – a Half-Century Survey*, ed. R. Herken, Oxford University Press, Oxford 1988, pp. 227–257.

[27] P. Clayton, *Mind and Emergence: From Quantum to Consciousness*, Oxford University Press, Oxford 2004, pp. 107–108.

inquiry? (b) Are mental/cognitive systems part of nature? (c) Does the mind have unique properties (functions) which do not appear at the level of microphysical, chemical, or biological systems? (d) Is the mind causally efficacious in the physical world? (e) Do socio-cultural factors play an important role in the development of the human mind? Psychophysical emergence is a special case of emergence, one that concerns the appearance of structures capable of realizing mental/cognitive functions. At least in their developed form, these functions cannot be found at the early stages of the evolutionary process, or deep in the microstructure of organisms. As Clayton aptly observes:

> Conscious states and experiences are not found in the individual neurons: they emerge out of the massively complex system that is the human brain. When we look at human persons and their qualities, we realize that the whole is indeed greater than the sum of the parts.[28]

The existence of minds like ours is only possible at a specific level of organization, in a relatively narrow section of the space-time continuum, in a Universe with suitably "selected" laws and initial conditions. Even a minor alteration of the laws or initial conditions would have made it impossible for any structures capable of realizing life and consciousness to develop. It is easy to imagine dead worlds, devoid of consciousness, without laws (programs) permitting sophisticated self-organization. Psychophysical emergentism is a view according to which human minds, as well as the cognitive structures of other animals, develop via evolutionary processes on the basis of neurobiological systems equipped with a brain and a central nervous system. In the case of human minds, the role of the environment and of historical, linguistic, and socio-cultural factors is equally important. The discovery of the structural, genetic, developmental, and functional dependence of mental processes on their neurocomputational

[28] *Ibid.*, p. 156.

base has rendered the Cartesian version of the mind-body problem obsolete. The mind is not a substance, but a process, a system of functions, relations, and activities or states generated by a working brain – or a physical structure functionally equivalent to a brain – embedded in a specific environment. What constitutes principal evidence for the emergence of mind from the brain are *strict dependencies* (they are no mere correlations!) between brain structures and mental functions. Damaging a brain structure underlying certain mental capacities leads to their degradation or destruction. This deep constitutive dependence is the subject of comprehensive empirical investigation and philosophical analyses.[29]

The laws of physics alone do not suffice to account for the structure of our brains and the mental/cognitive functions they perform. Furthermore, the laws that contribute to shaping the structure of the brain, e.g. the laws of genetics, do not follow directly from physical laws. The architecture of our brains depends to a large extent on random variables, like mutations and environmental factors. This is clear from the inter-individual and inter-specific differences between brains (minds) encountered in nature. No brain is exactly the same as any other brain, even within one species. Although neurophysiological processes are a necessary substrate of the mind, it is useful at a higher level of description, analysis, and explanation to posit an abstract mental space where mental events (sensations, conscious experiences, emotions, intentional contents, etc.) are situated. Mental

[29] R. Sperry, *Mental Phenomena as Causal Determinants in Brain Function*, [in:] *Consciousness and the Brain*, eds. G.G. Globus, G. Maxwell, I. Savodnik, Plenum Press, New York – London 1976, pp. 162–171; R. Sperry, *Mind-Brain Interaction: Mentalism, Yes; Dualism, No*, “Neuroscience” 1980, vol. 5, pp. 195–206; R. Sperry, *Macro- versus Micro-Reductionism*, “Philosophy of Science” 1986, vol. 53, pp. 265–270; M. Silberstein, *Emergence and the Mind-Body Problem*, “Journal of Consciousness Studies” 1998, vol. 5 (4), pp. 464–482; M. Silberstein, *Converging on Emergence*, “Journal of Consciousness Studies” 2001, vol. 8 (9–10), pp. 61–98; Ch. Koch, *The Quest for Consciousness: A Neurobiological Approach*, Roberts and Company Publishers, Engelwood 2004; J. Hattiangadi, *The Emergence of Minds in Space and Time*, [in:] *The Mind as a Scientific Object: Between Brain and Culture*, eds. C.E. Erneling, D.M. Johnson, Oxford University Press, Oxford 2005, pp. 71–100.

events enjoy *relative autonomy* at least in the sense that their specificity and characteristics are not identical with the properties of their neural substrates. The abstract mental space contains the reduced patterns (mental images, concepts, beliefs) of structures that are external to the cognitive system, as well as a representation of the body and a dynamic representation of the Self. We construct our inner, partly autonomous mental space by creating complex models and representations of the world.[30]

The mental patterns we become aware of are internal (subjective, individual) states of the mind that arise as a result of activity of neural structures in our brains. Relations between mental states, though constrained by the physical structures of the brain, form an emergent, partly autonomous inner world. Physical and neurobiological phenomena are essential to the emergence of mental phenomena, but they fail, in an important sense, to explain them. The world of the mind cannot be reduced completely to neural activity, just as cultural patterns cannot be reduced completely to the properties of a single mind.[31]

We no longer ask: Can a mental substance affect a mechanical brain? *Can* physical, chemical, biological, neural, and computational processes lead to the emergence of mind and consciousness? What we now ask is: How do mental states arise from lower-level (physical, chemical, neurobiological) processes? How are structures from the outside world mirrored in brain structures? How do cognitive systems represent the environment? What sort of mechanisms underlie conscious mental activity? These questions give rise to a multitude of more specific problems. It should come as no surprise, then, that so many scientific disciplines have been recruited to solve the mind-body problem. We are now seeing considerable convergence of research on various aspects of the multidimensional, emergent pheno-

[30] P. Gärdenfors, *Conceptual Spaces: The Geometry of Thought*, MIT Press, Cambridge, Mass. 2000; D. Hofstadter, *I Am a Strange Loop*, Basic Books, New York 2008.
[31] A. Scott, *Reductionism Revisited*, "Journal of Consciousness Studies" 2004, vol. 11 (2), pp. 51–68.

menon that is the human mind. Discovering the natural mechanisms of its emergence does not diminish the exceptional status of mankind in nature. It does change, however, our view of the nature of the Universe and of the potentialities it encodes.[32]

3. The Limits of Neuroscientific Explanations of Mind

3.1. Richard Feynman's remarks on levels of organization in nature are an interesting illustration of the emergentist line of thought. According to Feynman, the existing hierarchy of scientific disciplines suggests a picture of the world as a hierarchical super-system consisting of innumerable subsystems. As we move up the hierarchy of complexity, we go from elementary particles to atoms, molecules, cells, and multicellular structures. Moving on, we encounter phenomena like a muscle twitch or nerve impulse, which are enormously complicated and require complex organization. Higher still, peculiar objects like frog, leopard, and man appear. Humans, in turn, enter into complex systems of interrelations, producing further structures and levels of social, economic, and political organization, which then influence their own mental states. At the level of behaviour and interpersonal relations, new concepts arise, like political correctness, hope, beauty, or the moral good, which did not occur at the lower levels. According to Feynman, we have only recently learned to see the whole of this structure, encompassing a multitude of two-way connections between history and psychology, psychology and neurology, neurology and biology, and chemistry and physics. It would be a mistake to treat one end of this hierarchical system as the fundamental level to which everything else would reduce. Instead of seeking unified, ultimate reductive explanations, we should carry out research at both ends of the

[32] K.R. Popper, *The Open Universe: An Argument for Indeterminism*, Routledge, London–New York 1982; A. Damasio, *Self Comes to Mind: Constructing the Conscious Brain*, William Heinemann, London 2010.

hierarchy ("upward" and "downward"), as well as in the middle. That is the only way in which we can gradually understand this tremendous world of interconnected levels.[33]

Feynman's remarks also apply to the fields of neuroscience, which analyze various levels of organization in the nervous system – the molecules, synapses, nerve cells, layers, columns, and complex functional systems controlling an organism's physiological, cognitive, and behavioural processes. These investigations have been divided among many special disciplines – from biochemistry and molecular biology, through cytology and histology, neurophysiology and neuroanatomy, to behavioural neuroscience, neuropsychology, and cognitive neuroscience. In this case, too, a hierarchy of scientific disciplines corresponds to a natural hierarchy of systems of various orders. What remains controversial is: (a) How exactly do the levels of description, analysis, and explanation posited within neuroscience represent the actual levels of organization in the nervous system? (b) How do the mechanisms and processes occurring at the different levels of a system realize mental/cognitive functions and processes?[34]

There are two, seemingly conflicting, research strategies in the philosophy of cognitive neuroscience. One might be called a *descending* approach, the other – an *ascending* one. Proponents of the descending strategy, known as *microreductionists*, believe that in order to explain the mind, one must understand how the brain works. And to know *that*, one needs to understand the chemical structure as well as principles of operation of neurons, axons, intramembrane proteins, amino acids, etc. It is not clear, however, where this explanatory procedure should stop.[35] By contrast, proponents of the ascending strategy, known as *emergentists*, point to the essential limitations

---

[33] R. Feynman, *The Character of Physical Laws*, MIT Press, Cambridge, Mass. 1965, ch. 5.

[34] P.S. Churchland, T. Sejnowski, *The Computational Brain*, MIT Press, Cambridge, Mass. 1992, pp. 18–19.

[35] J. Bickle, *Philosophy and Neuroscience: A Ruthlessly Reductive Account*, Kluwer, Dordrecht 2003, p. 3.

of accounting for complex systemic functions in terms of the system's microstructure. They believe the reductionist claim – namely, that higher levels of organization in mental/cognitive systems (and their corresponding theories) can be reduced without remainder to the lower levels (and their corresponding theories) – to be untenable. If what appears at a higher level are qualitatively and functionally distinct units which display their own dynamics and are governed by their own laws, then their microreductive explanations must inevitably remain partial. This is especially true of the descriptions, analyses, and explanations of mental/cognitive processes and socio-cultural phenomena in terms of the concepts and theories of neurobiology. As Harold Morowitz rightly observes, despite the often-encountered tendency to treat the two approaches described as mutually incompatible, one can view them as complementary to each other:

> The reductionist approach leads us continually to seek solutions at lower and lower hierarchical levels. To move conceptually in the other direction, we must apply pruning algorithms and seek for emergent properties or entities that become the agents for advancing another hierarchical level. [...] Nature yields at every level novel structures and behaviours selected from the huge domain of the possible by pruning, which extracts the actual from the possible. [...] Emergence is then the opposite of reduction. The latter tries to move from the whole to the parts. It has been enormously successful. The former tries to generate the properties of the whole from an understanding of the parts. Both approaches can be mutually self-consistent.[36]

3.2. Contemporary philosophy of neuroscience is moving away from the deductive-nomological model of explanation and reduction toward the mechanistic model. It is pointed out that, rather than invoke

[36] H. Morowitz, *The Emergence of Everything: How the World Became Complex*, Oxford University Press, Oxford 2002, p. 14.

strictly general laws, neurocognitive accounts of visual perception, emotions, memory, and consciousness appeal to neurobiological and computational mechanisms operating at various levels of complexity in the nervous system. Mechanist explanations describe hierarchically organized mechanisms, each consisting of a multitude of integrated processes that realize relatively simple operations.[37] By combining these processes, it is possible to produce enormously complex functions, such as perception, memory, or consciousness – a fact that is often taken to support a version of the principle that a whole can be greater than the sum of its parts.[38]

In classical philosophy of science, the term "intertheoretic reduction" was used when the laws of a reduced theory were derived from the laws of a reducing theory in conjunction with what was known as bridge laws. On the mechanistic model, "reduction" has acquired a more ontic interpretation. For example, a reductive mechanist explanation of memory involves specifying a number of micromechanisms, occurring at various levels of organization in the nervous system, such that their joint operation makes it possible for complex memory processes to arise. Mechanistic reductions and explanations commonly rely on multiple neuroscientific theories, and can be accomplished on the basis of interlevel theories developed within interdisciplinary research programs. This kind of reduction is compatible with moderate models of emergence. On the mechanistic view, higher-order functions and processes are neither eliminated nor identified with their simple component operations. Rather, the aim is to show how complex systemic functions emerge out of a number of simpler mechanisms and operations occurring at many different levels of organization in the nervous system.[39]

---

[37] W. Bechtel, *Mental Mechanisms: Philosophical Perspectives on Cognitive Neuroscience*, Routledge, New York–London 2008, ch. 1, 2, 4, 7.
[38] C.F. Craver, *Explaining the Brain: Mechanisms and the Mosaic Unity of Neuroscience*, Clarendon Press, Oxford 2007, p. 189.
[39] C. Wright, W. Bechtel, *Mechanisms and Psychological Explanation*, [in:] *Handbook of the Philosophy of Science: Philosophy of Psychology and Cognitive Science*, ed. P. Thagard, Elsevier, Amsterdam 2007, pp. 59–60.

3.3. Mechanistic explanations of mental/cognitive processes do not always proceed by analyzing a given system into parts (structural decomposition) or component operations (functional decomposition). Another explanatory strategy that can be employed is what Carl Craver calls a *contextual explanation*. We adopt it when, in explaining a phenomenon at level *n*, we appeal to mechanisms and operations at level *n*, or *n* + 1.[40] Contextual explanations of mental processes go beyond purely neuroscientific explanations when they appeal to external determinants, such as social or environmental factors.

An important feature of natural systems is that they can affect the operation of their subsystems. A subsystem's properties may vary considerably depending on what kind of higher-order system it belongs to. And vice versa: some subsystems can regulate the activities of a whole system. What we see in such cases is a complicated network of feedback mechanisms known as upward and downward causation. The notion of downward causation is a definitional feature of an important class of concepts of emergence. Downward causation can be identified with what regulates the system, constrains it, or specifies the boundary conditions for the system's internal processes; it can also be understood as the sort of active agency that we encounter at the level of conscious intentional systems.[41] The character of downward causation depends on the kind of system involved. Some authors prefer to use the terms "systemic" or "contextual causation" instead of "downward causation".[42]

---

[40] C.F. Craver, *Explaining…*, *op. cit.*, p. 256.

[41] D.C. Campbell, *Levels of Organization, Downward Causation, and the Selection--Theory Approach to Evolutionary Epistemology*, [in:] *Theories of the Evolution of Knowing*, eds. G. Greenberg, E. Tobach, Erlbaum, Hillsdale 1990, pp. 1–17; C. Emmeche, S. Koppe, F. Stjernfelt, *Levels, Emergence, and Three Versions of Downward Causation*, [in:] *Downward Causation: Minds, Bodies and Matter*, eds. P.B. Andersen, C. Emmeche, N.O. Finnemann, P.V. Christiansen, Aarhus University Press, Aarhus 2000, pp. 13–34.

[42] M. Silberstein, *In Defense of Ontological Emergence and Mental Causation*, [in:] *The Re-Emergence of Emergence*, eds. P. Clayton, P. Davies, Oxford University Press, Oxford 2006, pp. 203–226.

Within systems theory, microreductive explanations are generally taken to be partial in character. This is also true of microreductive explanations in neuroscience. As a matter of fact, every system, with the possible exception of the universe itself, is a subsystem of some higher-order system – which means that it is subject not only to micro-macro, but also to macro-micro interactions (of the downward causation type). Even simple observations provide numerous illustrations of these macro-micro influences (when a limb is severed, its cells die; when the sun disappears beyond the horizon, the mean kinetic energy of the air molecules decreases; the divorce of a married couple has repercussions for other family members, etc.). In each case, the state or property of a given system changes as a result of what happens at the level of a higher-order system, of which the former system is a part. As Mario Bunge remarks:

> The failures of full micro-reduction may be explained by the hypothesis that every real thing, except for the universe as a whole, is embedded in some higher-level system [...]. In short, although reduction ought to be pushed as far as possible, we should realize its limits: we should settle for partial (or weak) reduction whenever full (or strong) reduction is unattainable. This methodological maxim distinguishes moderate from radical reductionism.[43]

3.4. On the contextual-emergence model, the properties determined by a system's surroundings are called its *relational emergent properties*.[44] Psychosemantic properties of beliefs, like truth and falsity, provide a case in point. According to psychosemantic realism, what determines the truth-value of a belief is not the microstructure of our

---

[43] M. Bunge, *Emergence and Convergence. Qualitative Novelty and the Unity of Knowledge*, University of Toronto Press, Toronto 2003, p. 148.

[44] M. Mahner, M. Bunge, *Foundations of Biophilosophy*, Springer, Berlin 1997, pp. 29–33; M. Silberstein, *op. cit.*.; H. Atmanspacher, *Contextual Emergence: From Physics to Cognitive Neuroscience*, "Journal of Consciousness Studies" 2007, vol. 14 (1–2), pp. 18–36.

brain, but some objective state of affairs in the world. Beliefs are not true or false because of how our brains are organized, but in virtue of a correspondence relation holding between an assertoric propositional content (the mental argument of the truth relation) and a state of affairs in the world (the extramental argument of the truth relation). The truth-values of beliefs do not supervene solely on the microstructure of a cognitive system that possesses them, because their determining factors reside in the outside world, not inside our heads (except *de se* beliefs). Having a suitably organized brain is a necessary condition of belief possession, but it is not a sufficient condition of the beliefs' truth. The psychosemantic aspect of the mind lies outside the purview of the explanatory procedures of neuroscience. Neuroscience will not displace psychosemantics, though it might shed light on how belief states are realized by neural mechanisms.[45]

An adequate model of the mind should take into account many levels of organization and their mutual interactions. According to Gerald Edelman, given the number of physical, biological, and social levels that must interact with one another in order for a conscious mind to arise, the idea of reducing the theory of mind to fundamental physics or some such low-level theory is downright silly. It is impossible to provide a comprehensive description and explanation of the human mind in terms of fields and particles, molecules, or even single neurons. Extreme versions of neuroreductionism fail because we know that, in order to operate properly, memory, higher-order consciousness, and the extended self need to connect to the level of social relations. A comprehensive description of such mental states as propositional attitudes cannot be achieved within the framework of a narrowly construed neurocomputational model, for it does not allow for reference to an external world. Given that beliefs and mental processes depend on social and cultural interactions, as well as on conventions, logic, and metaphors, any attempt to account for them

[45] R. Wilson, *Boundaries of the Mind: The Individual in the Fragile Sciences*, Cambridge University Press, Cambridge 2004, ch. 6–7.

by appeal to purely neurobiological methods must be unsatisfactory. A reduction of psychology and psychosemantics to neurobiology would be a partial reduction at best.[46]

Microreductive explanations, in neuroscience as well as everywhere else, are generally incomplete, which does not change the fact that they ground higher-level processes in lower-level mechanisms and structures. Attempts to replace higher-order theories by their lower-level counterparts always lead to a loss of information, since higher-order theories and disciplines are richer in information content. Chemistry provides informational enrichment to physics, biology enriches chemistry, psychology enriches neuroscience, and the social sciences enrich psychology. This enrichment is conceptual-informational in character and, in many cases, nomological (at the higher levels of organization we encounter novel regularities and new principles of organization which are epistemically inaccessible from the perspective of lower-order theories). The limits of neuroscientific explanations of mind are a special case of the limitations of microreductionist programs, which follow from the systemic nature of the world.

Conclusion

If our brains lacked the appropriate structure, they could not support parallel distributed processes of information processing and integration. As a result, we would be unable to realize even the most elementary mental activities. Perception, memory, inference, consciousness, decision-making, creative activity, along with scientific inquiry and the capacity to detect regularities in nature, would be impossible. The repertoire of the mental/cognitive states and activities we can realize is determined by the global architecture of our brains. We can

[46] G. Edelman, *Bright Air, Brilliant Fire: On the Matter of the Mind*, Basic Books, New York 1992, ch. 14, 16.

have only those mental/cognitive states that the neurodynamics of our brains permits us to have. This does not mean that models of the mind are identical with models of the brain. The former require a different level of analysis from the latter. An analysis of the mind at the level of its substrate, though extremely explanatorily fruitful, does not suffice. What is also necessary are detailed psychological, psychosemantical, and phenomenological analyses revealing higher levels of structural and functional organization (including the personal and social ones) in the cognitive system.

According to an often cited platitude, neurophysiological processes are an essential substrate of mental processes. We should neither overrate nor underrate this. What we need in order to perceive, think, speak, remember, and experience jealousy or annoyance is a modular brain characterized by a specific functional organization. Having such a brain is necessary, but not sufficient. For example, to explain an intentional feeling of jealousy it would not do to simply describe the activation pattern of particular cortical modules, the attendant changes in cerebral bloodflow, or the areas of increased oxygen consumption in particular brain regions; although, plausibly, some such distinctive patterns of activation will occur. A full account must also encompass a network of social and interpersonal relations in which a given person is involved. Even a complete knowledge of that person's brain will not give us an insight into the phenomenological or relational nature of the state of jealousy she is experiencing. What is needed are phenomenological descriptions and contextual analysis. By only studying the modular organization of the brain we would have no chance of learning a great many things. This is a fundamental barrier. Systemic research, treating the brain as a subsystem of many higher-level systems, is necessary.

We can achieve comprehensive explanations of the mind by looking at the brain and the holistically construed cognitive system as parts of a complex network of synchronic and diachronic, vertical and horizontal relations. It is impossible to capture the relational aspect of the mind exclusively in terms of its modular neurobiological substrate. Because a narrowly conceived modular theory of mind can merely hope to provide a partial account of the mechanisms underlying the mind, it cannot serve as the only theoretical background for a philosophy of mind. The philosopher's perspective, if he or she is after a comprehensive conception of the mind, must be broader: it must take into account the contribution of external determining factors, whether they be environmental, developmental, historical, socio--cultural, or logical and axiological in character.

Marcin Miłkowski
Polish Academy of Sciences

Theoretical Unification and the Neural Engineering Framework[1]

Introduction

The explanatory practices in neurosciences vary wildly, and there is a good reason for this: there are simply different theoretical interests in different branches of neural research. The chemical processes in individual neurons call for a different kind of explanation than processes explained by cognitive neuroscience. Yet in cognitive neuroscience – and in systems neuroscience – it is a common complaint that there aren't many theories, or any theories for that matter at all. The Neural Engineering Framework (NEF)[2] is one of the few approaches that has been defended as a general theory in this field. In this chapter, I deal with the question of whether the NEF is a unifying approach in the light of Philip Kitcher's unification account of explanation.[3] Elsewhere I have defended the view that the NEF is genuinely explanatory according to the principles of the mechanistic account of explanation[4];

[1] This paper was written within the research grant "The Limits of Scientific Explanation" sponsored by the John Templeton Foundation. The author wishes to thank Chris Eliasmith for helpful comments to the earlier version of this chapter.

[2] C. Eliasmith, C.H. Anderson, *Neural Engineering. Computation, Representation, and Dynamics in Neurobiological Systems*, MIT Press, Cambridge, Mass. 2003.

[3] P. Kitcher, *Explanatory Unification and the Causal Structure of the World*, [in:] *Scientific Explanation*, eds. P. Kitcher, W.C. Salmon, no. 505, University of Minnesota Press, Minneapolis 1989, pp. 410–505.

[4] M. Miłkowski, *Explaning the Computational Mind*, MIT Press, Cambridge, Mass. forthcoming. For a general exposition of neo-mechanist philosophy of science, see

so the question is not whether the NEF is explanatory at all but rather whether it exhibits the properties of theories that are best characterized as involving theoretical unification.

I start with the general exposition of the idea of theoretical unification. Then I present the general principles of the NEF and introduce two example explanations: one related to the navigational capabilities of rats,[5] and one to the human performance during the Wason task.[6] The proponents of the NEF stress that it has the feature of unifying previously disparate phenomena by offering a common methodology but do not analyze it in Kitcher's terms. It is therefore only natural to see whether there is a common argument pattern in the NEF explanations – something that Kitcher thinks is the crucial ingredient of unification.

1. Theoretical Unification

According to Philip Kitcher, there are two basic ways of dealing with the objections to the received view of explanation in the philosophy of science, which is the covering-law (CL) account developed by Carl Hempel and Paul Oppenheim.[7] Namely, one can either dispose of the

---

C.F. Craver, *Explaining the Brain. Mechanisms and the Mosaic Unity of Neuroscience*, Oxford University Press, Oxford 2007; P. Machamer, L. Darden, C.F. Craver, *Thinking about Mechanisms*, "Philosophy of Science" 2000, vol. 67, no. 1, pp. 1–25.

[5] J. Conklin, C. Eliasmith, *A Controlled Attractor Network Model of Path Integration in the Rat*, "Journal of computational neuroscience" 2005, vol. 18, no. 2, pp. 183–203.

[6] C. Eliasmith, *Cognition with Neurons: A Large-scale, Biologically Realistic Model of the Wason Task*, [in:] *Proceedings of the XXVII Annual Conference of the Cognitive Science Society*, eds. G. Bara, L. Barsalou, M. Bucciarelli, Stressa 2005; C. Eliasmith, *Learning Context Sensitive Logical Inference in a Neurobiological Simulation*, [in:] *Compositional Connectionism in Cognitive Science*, eds. S. Levy, R. Gayler, AAAI Press, 2004, pp. 17–20; P.C. Wason, *Reasoning*, [in:] *New Horizons in Psychology*, ed. B.M. Foss, Penguin, Harmondsworth 1966, pp. 135–151.

[7] C.G. Hempel, P. Oppenheim, *The Logic of Explanation*, [in:] *Readings in the Philosophy of Science*, eds. H. Feigl, M. Brodbeck, Appleton-Century-Crofts, Inc., New York 1953, pp. 319–352.

argument-like structure as essential to scientific explanation, and say that the crucial explanatory relationship is causal in nature, or retain the basic argumentative account of explanation, but eliminate the problematic reference to the notion of the law.[8] Other ways – such as adding additional logical or pragmatic requirements to the Hempelian view – seem deeply unsatisfactory to him. As well, he finds the causal conception problematic, especially when applied to formal sciences. These are all contentious claims, and I will not discuss them here. What is crucial is that the unification account is supposed to be the improved version of the CL framework.

According to the CL framework, an explanation is an argument whose premises (the *explanans*) contain at least one lawlike statement, and whose conclusion describes the *explanandum* phenomenon. In the case of deductive arguments whose conclusions are singular statements, the explanations are of deductive-nomological (D-N) kind; in case of statistical or inductive ones, the relationship between the premises and the conclusion is not deductive. Kitcher generalizes the CL account by no longer requiring that the law is among the premises. This might strike one initially as a radically erroneous: if laws are not required, then why should one call this explanation at all? The answer is that the unification account has more on offer. Namely, it can deal with sciences (such as evolutionary biology) where universal laws, required by Hempel, are scarce or non-existent. Rather, one can find mini-laws of a limited scope, or simply limited invariant generalizations.

The main idea of unification is that if we can explain much with a little, then the explanation is genuine.[9] So if a theory can reduce the number of allegedly independent phenomena by subsuming them under a single argument pattern – note: not under a single law – and thus get rid of many independent lawlike or invariant generalizations, then it is explanatory:

---

[8] P. Kitcher, *Explanatory...*, *op. cit.*

[9] Kitcher draws from an earlier, though largely failed account of unification: M. Friedman, *Explanation and Scientific Understanding*, "The Journal of Philosophy" 1974, vol. 71, no. 1, pp. 5–19.

> Science advances our understanding of nature by showing us how to derive descriptions of many phenomena, using the same patterns of derivation again and again, and, in demonstrating this, it teaches us how to reduce the number of types of facts we have to accept as ultimate (or brute).[10]

The explanations earn their title because they belong to the *explanatory store* E, and "the fundamental task of a theory of explanation is to specify the conditions on the explanatory store".[11] The set of statements endorsed by a scientific community at a given time – K – needs to be explained, and the set of derivations, according to the unification account, that best systematizes K is E(K). As Kitcher claims, "E(K) is a set of derivations that makes the best tradeoff between minimizing the number of patterns of derivation employed and maximizing the number of conclusions generated".[12]

Unification is explicated in terms of common – or similar, as Kitcher allows for similarity of derivations as well – argument patterns: "a *general argument pattern* is a triple consisting of a schematic argument, a set of sets of filling instructions, one for each term of the schematic argument, and a classification for the schematic argument," where the classification is "a set of statements describing the inferential characteristics of the schematic argument: it tells us which terms of the sequence are to be regarded as premises, which are inferred from which, what rules of inference are used, and so forth".[13] Instead of analyzing these definitions, let me introduce an example that Kitcher himself proposed as a simplified version of Mendelian genetics:

(1) There are two alleles *A*, *a*. *A* is dominant, *a* recessive.

(2) *AA* (and Ad) individuals have trait *P*, *aa* individuals have trait *P'*.

[10] P. Kitcher, *Explanatory...*, *op. cit.*, p. 432. Italics removed.

[11] *Ibid.*, p. 430.

[12] *Ibid.*, p. 432.

[13] *Ibid.*

(3) The genotypes of the individuals in the pedigree are as follows: i_1 is G_1, i_2 is G_2, . . . , i_N is G_N. {(3) is accompanied by a demonstration that (2) and (3) are consistent with the phenotypic ascriptions in the pedigree}.

(4) For any individual *x* and any alleles *yz* if *x* has *yz* then the probability that *x* will transmit *y* to any one of its offspring is 1/2.

(5) The expected distribution of progeny genotypes in a cross between i_j, and i_k is *D*; the expected distribution of progeny genotypes in a cross . . . {continued for all pairs for which crosses occur}.

(6) The expected distribution of progeny phenotypes in a cross between i_j and i_k is *E;* the expected distribution of progeny phenotypes in a cross . . . {continued for all pairs in which crosses occur}.

Filling Instructions: *A*, *a* are to be replaced with names of alleles, *P*, *P'* are to be replaced with names of phenotypic traits, i_1, i_2, . . . , i_N are to be replaced with names of individuals in the pedigree, G_1, G_2, . . . , G_N are to be replaced with names of allelic combinations (e.g. *AA*, *Aa*, or *aa*), *D* is replaced with an explicit characterization of a function that assigns relative frequencies to genotypes (allelic combinations), and *E* is to be replaced with an explicit characterization of a function that assigns relative frequencies to phenotypes. *Classification*: (1), (2), and (3) are premises; the demonstration appended to (3) proceeds by showing that, for each individual *i* in the pedigree, the phenotype assigned to *i* by the conjunction of (2) and (3) is that assigned in the pedigree; (4) is a premise; (5) is obtained from (3) and (4) using the principles of probability; (6) is derived from (5) and (2).[14]

Even if in Mendelian genetics there were 'laws' that were supposed to support the explanations, Kitcher analyzes the example without any appeal to the notion of law. This is, of course, in line with the later developments in genetics, when it turned out that these generalizations

[14] *Ibid.*, p. 439.

had to be hedged with multiple conditions and exceptions. It is arguable that such a reconstruction of Mendelian genetics actually works better than the traditional CL account, as we no longer need to worry about the lawlike character of generalizations.

Kitcher argues that this version of argument-like conception of explanation deals successfully with two main kinds of objections against the CL account: the irrelevant factors objection, and the asymmetry objection. The irrelevant factors objection is related to the fact that adding a true sentence to the premise set and to the conclusion will not make the conclusion false and the validity will be preserved. If the original explanation of the fact that a sample S of table salt dissolved in water was derived from the law that all table salt dissolves in water, and that sample S is a sample of table salt, then we can as easily derive the conclusion that a sample S of hexed table salt dissolved in water from the premises that a sample S was a sample of hexed table salt, and that all hexed table salt dissolves in water. Of course, the fact that the sample was hexed by a magician is irrelevant. Note that you cannot exclude this counterexample to Hempel's account by requiring that the premise set needs to be minimal with respect to the explanandum, as the explanandum itself already includes hexed salt. The solution that Kitcher recommends is straightforward: the argument pattern that includes the 'hexing' predicate will not be useful in other derivations in chemistry. Either it will be an idle factor, which is a candidate for elimination because of its irrelevance in descriptions, or it will not be a part of a genuine projectible predicate. The problem for Kitcher, who relies on Goodman's idea of projectibility,[15] is that Goodmanesque criteria are prone to simple and disastrous objections.[16] One answer to this worry would be to say that all that is required by Kitcher is that there was at least one way of saying when a predicate is projectible,

[15] N. Goodman, *Fact, Fiction, and Forecast*, 4th ed., Harvard University Press, Cambridge, Mass. and London 1983.

[16] A. Zabludowski, *Concerning a Fiction about how Facts are Forecast*, "The Journal of Philosophy" 1974, vol. 71, no. 4, pp. 97–112.

and when it is not; it need not be the original Goodman solution to this problem. I assume that a proponent of the unification theory of explanation could appeal to the fact that the information complexity of theories containing gerrymandered predicates, such as "hexed or unhexed and composed of NaCl", would be simply larger than of the theory that does not contain them. For example, by using minimal-length complexity, or Kolmogorov complexity, one may easily compare theories this way.[17]

The asymmetry objection concerns the problem that one can replace a premise with a conclusion in a sound argument and retain soundness. For example, suppose that I want to explain the fact that a tower has a shadow of length of 16.78 m by citing four premises: (a) any tower of a height X when lit with Sun light from the apparent altitude of Y will be (insert a formula here);[18] (b) the tower has height of 20 m; (c) the altitude is 50. But the argument stays equally fine, according to the CL account, when the conclusion is that the tower has height of 20 m which is inferred from facts about its shadow and the Sun's apparent altitude. But surely that is not explanatory! Shadows are not relevant factors in laws of nature; they appear because of the objects' occlusion of the Sun light. Kitcher's response is rather lengthy, but instead of focusing on a simple example above, he seems to be distracted by an answer to this problem suggested earlier by van Fraassen, who claimed that in some contexts shadow lengths could be explanatory of heights (think of solar clocks, for example). Accordingly, Kitcher focuses on origins and development of entities as explanatory factors for their properties, because some entities, like in van Fraassen's example[19], were intended to cast some shadow. These

---

[17] For introduction to Kolmogorov complexity and minimum description length, see: M. Li, P. Vitanyi, *An Introduction to Kolmogorov Complexity and Its Applications*, Springer-Verlag, New York–Berlin–Heidelberg 1993; P.D. Grünwald, *The Minimum Description Length Principle*, MIT Press, Cambridge, Mass. 2007.

[18] Needless to say, I ignore many important factors, such as the location of the tower on Earth, the time of year etc. But the simplistic version will suffice.

[19] B. van Fraassen, *The Scientific Image*, Clarendon Press, Oxford 1980, pp. 132–4.

cases are indeed solvable using unification criteria (the details need not interest us here):

> Explanation proceeds by tracing the less fundamental properties of things to more fundamental features, and the criterion for distinguishing the less from the more fundamental is that appeal to the latter can be made on a broader scale. Thus an attempt to subvert the order of explanation shows up in the provision of an impoverished set of derivations…[20]

So the response is based on the idea that the scope of explanation is negatively correlated with its being genuine. The argument pattern that allows inferring heights from shadows enables less true derivations than the inverse one, just because some objects do not cast shadows (being perfectly transparent, emitting light etc.).[21] Yet this is not a good answer to my simplistic counterexample. The premises in the example cited remain true after inversion, and the argument patterns are equally applicable, *pace* Kitcher. The conclusion in the first argument would be untrue if there were no shadow either, so it hard to say why should the first argument be an instantiation of a correct pattern, and the other not. It might as well turn out that Kitcher's proposal fails to answer this objection just like the traditional CL account.[22]

Where does this leave us? Apparently, the claims of the unification account seem to be blown out of proportion. The conception does not deal with simple counterexamples in a satisfactory way. However, this does not mean that theoretical unification is without merit. The claim that it is a virtue of scientific theories to have a wide scope of application while remaining also informative is uncontested. My

---

[20] P. Kitcher, *Explanatory…*, *op. cit.*, p. 487.

[21] Obviously, in both cases the set of derivations is infinite, which is another problem for Kitcher's account in general. It seems that it would have been easier to talk of the descriptive scope of theories not in terms of derivations but in terms of models (as the semantic view of theories would recommend).

[22] E. Barnes, *Explanatory Unification and the Problem of Asymmetry*, "Philosophy of Science" 1992, vol. 59, no. 4, pp. 558–571.

use of Kitcher's theory will be therefore instrumental: instead of using Kitcher's theory for testing whether the NEF is explanatory[23] or not, I will ask whether the NEF offers genuine unification in neurosciences.

2. The Neural Engineering Framework

In modern computational neuroscience, a range of methods and mathematical techniques are used to describe neural systems,[24] and it is difficult to talk of a single overarching methodology or theory.[25] The view that there aren't any good theories in neuroscience is quite popular.[26] The Neural Engineering Framework (NEF) is supposed to fill this theoretical void by supplying a general methodology that will unify neuroscientific explanations.

The NEF is not a theory of the neural system at its molecular level of organization and it is supposed to deal with one very important kind of explanations in systems neuroscience, namely representational explanation. Namely, neural representations are used to explain behavioural capacities. Interestingly enough, they are also explananda, i.e., it is explained how neural populations come to function as representations. All traditional representational explanations in cognitive science and neuroscience tackle the question: how does the input information get transformed to get output information? However, different frameworks construe the basic notions in different ways. What is especially pressing in this context is that many assumptions concerning the models built to explain the

[23] This case has been already made by C. Eliasmith, and it can be rephrased in mechanistic terms, as I previously did. M. Miłkowski, *Explaining…*, *op. cit.*

[24] Cf. for example P. Dayan and L.F. Abbott, *Theoretical Neuroscience. Computational and Mathematical Modeling of Neural Systems*, MIT Press, Cambridge, Mass. 2001.

[25] C. Eliasmith, *Neurocomputational Models: Theory, Application, Philosophical Consequences*, [in:] *The Oxford Handbook of Philosophy and Neuroscience*, ed. John Bickle, Oxford University Press, Oxford 2009, pp. 346–369.

[26] See e.g. C. Eliasmith, C.H. Anderson, *Neural…*, *op. cit.*, pp. 23–4 for many references.

representational function of the neural system seem to be ad hoc or merely technical. In other words, they are just needed to make the modelling work, but their theoretical status is dubious. This is in particular true of the choice of representational formats and their particular implementations.

The NEF is supposed to supply three basic principles that will help build representational explanations in a systematic way. The basic principles are:

(1) neural *representations* are understood as combinations of nonlinear encoding and optimal linear decoding (this includes temporal and population representations);

(2) *transformations* of neural representations are functions of variables represented by a population;

(3) neural *dynamics* is described with neural representations as control theoretic state variables.

These principles are also formulated in a strict mathematical form,[27] but for our purposes the informal version will suffice. Before I go on to explicate the basic ingredients of the NEF, it is important to note the additional fourth assumption contained in the framework, namely that noise is prevalent in biological systems; hence, any biologically realistic analysis must account for the effects of noise, and show how stability in the face of noise is achieved.[28]

Encoding and decoding are understood in engineering terms (hence *engineering* in the name of the framework): simply, the incoming information is encoded, just like in an artificial analog-to-digital device, and then remains decodable by the neural system that uses it. Obviously, the decoding does not serve the purpose of restoring the input signal (why bother with encoding it, were it possible to use the input signal directly) but of using the information that was contained in it. The input signals are encoded in neural populations (which is the main unit of analysis, although populations are comprised of single cells). In

[27] For details, see: *ibid.*, pp. 230–1.

[28] See *ibid.*, p. 16 and sections 2.2, 5.2.

their introductory textbook, Eliasmith and Anderson show three example representation formats. These are representations of scalars, vectors, and functions; and in certain – more speculative – extensions of the NEF, populations also encode structured representations needed to tackle compositionality-related problems.[29] All representations are analyzed in a consistent fashion in terms of nonlinear encoding of information and optimal linear decoding realized by temporal activity of neuron populations. In other words, when representing a variable, the system's representation is characterized by determining how to decode the value of that variable encoded into neural spikes trains.

Notably, the neuron populations are not *trained,* as in many other neural modelling frameworks, but analytically derived from basic equations. This shows both that the framework offers sufficient level of detail to be able to replace ad hoc modelling assumptions; for practical purposes, it means that the network training phase of modelling can be skipped, which saves researchers' time.[30]

Transformation in the NEF is synonymous with computation (in the sense employed in computability theory); however, as the latter term is sometimes understood in a fairly idiosyncratic way in the cognitive science, along the lines of "rule-governed discrete symbol manipulation",[31] a more neutral term "transformation" was chosen, which is understood along the same lines as representation; it "consists

---

[29] T.C. Stewart, T. Bekolay, C. Eliasmith, *Neural Representations of Compositional Structures: Representing and Manipulating Vector Spaces with Spiking Neurons*, "Connection Science" 2011, vol. 23, no. 2, pp. 145–153. See also C. Eliasmith, *Cognition with Neurons: A Large-scale, Biologically Realistic Model of the Wason Task*, *op. cit.* and T.C. Stewart, C. Eliasmith, *Building Production Systems with Realistic Spiking Neurons*, [in:] *Proceedings of the 30th Annual Meeting of the Cognitive Science Society*, eds. B.C. Love, K. McRae, V.M. Sloutsky, Cognitive Science Society, Austin, TX 2008, pp. 1759–1764.

[30] However, where learning is important for explanation, it is modelled as well, see e. g. D. MacNeil, C. Eliasmith, *Fine-Tuning and the Stability of Recurrent Neural Networks*, "PLoS ONE" 2011, vol. 6, no. 9, p. e22885; T. Stewart, T. Bekolay, C. Eliasmith, *Neural Representations of Compositional Structures: Representing and Manipulating Vector Spaces with Spiking Neurons,* "Connection Science", vol. 23, no. 2, pp. 145–153.

[31] See for example T. Van Gelder, *What Might Cognition Be, If Not Computation?*, "The Journal of Philosophy" 1995, vol. 92, no. 7, pp. 345–381.

of estimating arbitrary linear or nonlinear functions of the encoded variable".[32] The decoders that compute any function of the encoded input, not just the identity function, are called *transformational decoders*. The account of representation and transformation in terms of nonlinearities is successful because of the nonlinear nature of neural encoding of information.

The framework, though it features certain theoretical entities, such as decoders, which are not measurable directly, shows significant predictive success. According to Eliasmith, the theoretical entities are justifiable because of the critical role they play in determining connection weights in the NEF models. Connection weights *are* measurable. More importantly, these weights determine the tuning sensitivities of individual cells, dynamic properties of their responses, and overall network behaviour. Each of these is measurable, and has been explained and predicted by the NEF models across many neural systems.[33]

Even if the NEF is similar to traditional cognitive neuroscience, where cognition is construed as transformation of representation, it also includes control theory, and Eliasmith emphasizes that it is crucial that representations play the role of control-theoretic state variables. This is of course in no way different from a number of other theories in contemporary cognitive neuroscience; however, it is opposed to traditional cognitive science in some ways. As Eliasmith sees it, only with control theory is realistic decomposition of the system into

[32] C. Eliasmith, *Neurocomputational Models: Theory, Application, Philosophical Consequences*, *op. cit.*, p. 354.

[33] P.D. Kuo, C. Eliasmith, *Integrating Behavioral and Neural Data in a Model of Zebrafish Network Interaction*, "Biological Cybernetics" 2005, vol. 93, no. 3, pp. 178–87; R. Singh, C. Eliasmith, *Higher-dimensional Neurons Explain the Tuning and Dynamics of Working Memory Cells*, "The Journal of Neuroscience: The Official Journal of the Society for Neuroscience" 2006, vol. 26, no. 14, pp. 3667–78; B.J. Fischer, J.L. Peña, M. Konishi, *Emergence of Multiplicative Auditory Responses in the Midbrain of the Barn Owl*, "Journal of Neurophysiology" 2007, vol. 98, no. 3, pp. 1181–93; A. Litt, C. Eliasmith, P. Thagard, *Neural Affective Decision Theory: Choices, Brains, and Emotions*, "Cognitive Systems Research" 2008, vol. 9, no. 4, pp. 252–273; D. MacNeil, C.Eliasmith, *Fine-Tuning and the Stability of Recurrent Neural Networks*, *op. cit.*

its explanatorily relevant component parts possible.[34] The advantage of control theory is that it allows the inclusion of a real-time dimension in the dynamical description of the neural system in a fairly easy way; and the state variables can be easily mapped onto the underlying physical systems given the other principles – in contrast to typical dynamical systems approaches to cognition. Notably, the decompositions can be understood as relating to parts of mechanisms, which is in line with the neo-mechanistic principles of explanation.

Before we proceed to the discussion of two example explanations in the NEF, it is useful to discuss the question in what way it might be understood as offering unification. Eliasmith and Anderson argue that being able to unify phenomena is one of the four features of the NEF that make it not only a theoretical framework but a full-blown theory of the neural system:

> Consider Newton's laws of motion. Newton's laws constitute a theory of motion for a number of reasons. First, they *unify* what was, until Newton, considered two distinct kinds of motion; celestial and sublunar.[35]

Other features mentioned in this context, in case you are wondering, are that laws are *data-driven* (have empirical content), that they are *useful* in prediction, and that they are *quantitative*. Needless to say, the NEF, just like Newton's theory, has all four properties.

But is this sense of "unifying" compatible with Kitcher's theory? It seems that it is in a straightforward way the same notion: instead of offering many different explanations for similar phenomena, the NEF is supposed to supply the same scheme for all of them. In particular, population and temporal representation, usually considered as separate, are subsumed under the single pattern; they are explained in

[34] C. Eliasmith, *How We Ought to Describe Computation in the Brain*, "Studies In History and Philosophy of Science Part A" 2010, vol. 41, no. 3, pp. 313–320.
[35] C. Eliasmith, C.H. Anderson, *Neural Engineering. Computation, Representation, and Dynamics in Neurobiological Systems*, *op. cit.*, p. 24.

terms of nonlinear encoding and optimal linear decoding. Also, three principles unify representation and computation, as well as dynamics of the neural population: they are explained as parts of a coherent control-theoretic structure. The simplistic informal version of the pattern could be:

(1) A neural population N encodes information I as a nonlinear encoding E in its temporal activity.

(2) The encoding E can be decoded by an optimal decoder D according to formula G, which is embedded in the synaptic weights between neighbouring neurons in N.

(3) The encoded information is a variable that is control-theoretic, and state-space evolution of N is described with a formula F.

(4) If N encodes information as E and decodes as D in a way described by G and its behaviour is governed by F, it has capacity C.

(5) N has a capacity C.

Filling Instructions. E and D are to be replaced with a derived quantitative description of the encoding and decoding; N is to be replaced with an empirically valid description of the neural population; G and F are to be replaced with empirically sound quantitative formulae that describe known results about N. C is to be replaced with a description of the capacity that specifies whether the capacity is representational (where the input information is to be retained in the variable represented by N) or computational (where the function of the input information is important).

Classification: (1)–(4) are premises; (5) is a conclusion.

In numerous examples in the NEF introductory textbook, it is shown how to derive appropriate formulas needed in this argument pattern, which remains sketchy without the mathematical content. But it seems already quite cogent to say that the NEF is actually unifying in Kitcher's sense. However, it may be asked if it is unifying systems neuroscience in a sufficiently deep way. Of course, this is a priori hard to resolve; but a short comparison with another recent efforts at unification in neuroscience might help to ascertain whether it is the case. A book with a subtitle "a unifying computational neuroscience approach"[36] – the theory offered by Edmund Rolls – seems to be a plausible candidate for comparison. And indeed, Rolls mentions unification in a couple of places, notably in the ending of the book:

> In the senses that there are a set of memory-related operations that are involved in all these types of processing, and that there are interactions between the systems involved, we now have a **unifying** approach to memory, attention, decision-making, and perception (including vision, taste and olfaction) in which computational neuroscience helps to provide a framework in which all the essential empirical findings from neuroscience can be linked into a coherent understanding.[37]

A similar but a more detailed explication of unification can be also found in the conclusions of Chapter 4:

> The approach described in this chapter is **unifying** in a number of ways. First, a set of simple organizational principles involving a hierarchy of cortical areas with convergence from stage to stage, and competitive learning using a modified associative learning rule with a short-term memory trace of preceding neuronal activity, provide a basis for understanding much processing in the ventral visual stream, from V1 to the inferior temporal visual cortex. Second, the same principles help to understand some of the processing in the dorsal visual stream by which invariant representations of the global motion of objects may be formed. Third, the same principles continued from the ventral visual stream onwards to the hippocampus help to show how spatial view and place representations

[36] E.T. Rolls, *Memory, Attention, and Decision-Making*, Oxford University Press, Oxford 2007.

[37] *Ibid.*, p. 530. Emphasis mine.

> may be built from the visual input. Fourth, in all these cases, the learning is possible because the system is able to extract invariant representations because it can utilize the spatio-temporal continuities and statistics in the world that help to define objects, moving objects, and spatial scenes. Fifth, a great simplification and economy in terms of brain design is that the computational principles need not be different in each of the cortical areas in these hierarchical systems, for some of the important properties of the processing in these systems to be performed.[38]

In the latter case, it transpires that Rolls' theory is much more detailed than the NEF, which is an overarching methodology for neuroscience research also for non-human organisms, where memory and decision-making might be realized in simpler ways. The second quote also provides details that could be used to write an argument pattern in Kitcher's sense.

The question what level of detail is needed in one's explanation cannot be answered without knowing the scope of application of the explanation. Depending on theoretical interests, one may be interested in general principles of different neuronal architectures, or in the human brain, or in any individual neural system, for that matter. These are all valid interests. Many frameworks that are described as unifying might be quite different in the level of detail they offer. There is also a trade-off between generality and depth, and the best theories should be both deep and general. But how could we tell if the price for unification is superficiality of the NEF? To see if this is the case, let us analyze some examples.

---

[38] *Ibid.*, p. 374. Emphasis mine.

2.1. Rats in the Dark

In this section, the model of the path integration used by the rat navigation system will be sketched.[39] It belongs to the broad class of models of animal navigation based on cognitive maps. Modern neuroscience vindicates the idea of cognitive maps posited in rats by Edward Tolman,[40] especially since they were located it in the hippocampus.[41] This is not to say that the hypothesis is uncontested; some remain somewhat unimpressed, claiming that the neural representations found in rats are not really maps.[42] All in all, models of navigation, including those of path integration in rats, abound in computational neuroscience, as the behavioural and neuroscientific evidence is rich.[43]

The rat is one of the few animals capable of dead reckoning, or path integration, as this navigational capacity is called in biology (it was Charles Darwin in 1873 who hypothesized that animals have inertial navigational systems).[44] Rats are known to be able to navigate in their environment even in complete darkness. For example, they can return to their starting location after exploring the environment, and their route back might be quite different from the path of their exploration (they need not retrace their own steps). The only cues that they

[39] J. Conklin, C. Eliasmith, *A Controlled...*, *op. cit.* For a less technical description, see C. Eliasmith, *Neurocomputational Models: Theory, Application, Philosophical Consequences, op. cit.*

[40] E.C. Tolman, *Cognitive Maps in Rats and Men*, "Psychological Review" 1948, vol. 55, no. 4, pp. 189–208.

[41] J. O'Keefe, L. Nadel, *The Hippocampus as a Cognitive Map*, Clarendon Press, Oxford University Press, Oxford, New York 1978.

[42] S. Benhamou, *No Evidence for Cognitive Mapping in Rats*, "Animal Behaviour" 1996, no. 52, pp. 201–212.

[43] O. Trullier et al., *Biologically Based Artificial Navigation Systems: Review and Prospects*, "Progress in Neurobiology" 1997, vol. 51, no. 5, pp. 483–544; A.D. Redish, *Beyond the Cognitive Map: From Place Cells to Episodic Memory*, MIT Press, Cambridge, Mass. 1999; B.L. McNaughton et al., *Path Integration and the Neural Basis of the "Cognitive Map"*, "Nature Reviews. Neuroscience" 2006, vol. 7, no. 8, pp. 663–78.

[44] For a modern summary of findings on path integration, see S.J. Shettleworth, *Cognition, Evolution, and Behavior*, Oxford University Press, Oxford, New York, 2010.

need for this are their own movements, hence it is possible for them to navigate in the dark. The widely accepted hypothesis is that the rat represents the environment on a mental (or neural) map and updates its location on it.

Evidence shows that the representation of the rat's location can be thought of as a bump of neural activity centred on the rat's current location estimate. To explain path integration at a neural level, Conklin and Eliasmith take into account the known physiological and anatomical properties of the neuron populations in the areas active during navigation. Then, they derive appropriate formulas to determine the organization that would implement a mechanism to reproduce the observed behaviour. The bump of activity is represented in their model as a two-dimensional bell-shaped function. The nonlinear encoding, according to the first principle of the NEF, is then posited (again, the details are not relevant here) based on the response of neurons observed in the active areas.

The next step in modelling is sketching a high-level control mechanism responsible for navigation to the starting location based only on self-motion velocity commands from the vestibular system. There is a stable function attractor that holds the bump of activity in the same location in the case of no movement or moves the bump in the case of self-movement: so the representation encodes the position in the environment and it is updated based on velocity commands from the motor system in two dimensions. It is only natural to model the attractor by using a continuous attractor network.[45] This transformation of the neural representation in the attractor network is biologically realistic, even if it is based on the leaky-and-integrate-and fire (LIF) neurons (as Eliasmith and Anderson stress, more biologically faithful neuron models can be used in the NEF, but the price is paid in terms of the

[45] For more on attractor networks, see C. Eliasmith, *Attractor Network*, "Scholarpedia" 2007, vol. 2, no. 10, p. 1380. See also C. Eliasmith, *A Unified Approach to Building and Controlling Spiking Attractor Networks.*, "Neural Computation" 2005, vol. 17, no. 6, pp. 1276–1314.

processing time[46]). It is quite clear that the model is an idealization, even if a biologically-realistic one. The representation is embedded into the postulated control system according to the third principle of the NEF: it is a state variable of the system.

Conklin and Eliasmith applied the model to simulate a four thousand-neuron large network. When the rat is completing a circular path, there is a drift error, namely 11%. However, it is much smaller than in previous models using three hundred thousand neurons[47] – in the latter case, the error was 100%! The available data concerning phasic phenomena in rats can be replicated. Three predictions are made by the NEF model of path integration: (1) cells will be velocity- (and position-) sensitive; (2) navigation system cells will have the same location in different environments; (3) the direction of the rat's head will *not* affect its ability to come back to the initial location. The third prediction contradicts earlier models of this phenomenon. In general, the model predicts facts at the behavioural level – concerning head direction and integration bias, and at the neural one – predicting a specific correlation between neuron tuning curves and the rat's environment.

It is pretty clear now that this explanation supplies details that may be used to fill the argument pattern above. The capacity C is, quite clearly, the ability to integrate the path in the environment; a four thousand neuron population is used as N, and the original paper contains exact formulas to be inserted as E, D, and F.

[46] See C. Eliasmith, C.H. Anderson, *Neural…*, *op. cit.*, p. 127 for a comparison of information transmission characteristics of various model neurons.

[47] A. Samsonovich, B.L. McNaughton, *Path Integration and Cognitive Mapping in a Continuous Attractor Neural Network Model*, "The Journal of Neuroscience: The Official Journal of the Society for Neuroscience" 1997, vol. 17, no. 15, pp. 5900–20.

2.2. Pick a Card

Also, the NEF seems to bridge the gulf between the low-level neuronal activity and high-level cognitive mechanisms. As I have already mentioned, it was used to replicate famous results of the Wason task in cognitive psychology. The Wason selection task involves a deficit – or an alleged deficit – in human reasoning that seems to be particularly difficult to explain. In its original formulation, the task consisted in answering the question what should be done to test whether a given conditional is true. On the table, there were four cards with numbers (odd or even) on one side, and letters (vowels or consonants). Half of them are turned so that numbers are visible, and half on their letter side. Imagine that there are cards with 1, 4, A, and C. The proposition to test is: "if a card shows an even number on one face, then its opposite face shows a consonant". The conditional in classical logic is false just in case the antecedent is true and the consequent false; so we should check whether a card with an even number does not have a consonant. This might be true for the card with "4" and "A" (you have to check both cards, and checking 1 or C is inappropriate). Most people do not turn these cards, however. But as evolutionary psychologists found out, in the social contexts, where the conditional expresses, for example, a rule such as "If you are drinking alcohol then you must be over 18", the "error" disappears and most people get it right.[48] The overall structure of the task was nevertheless questioned by logicians, as it is based on the interpretation of "if" as a material implication, which is definitely not unquestionable in natural language. So while the experimental results

[48] L. Cosmides, *The Logic of Social Exchange: Has Natural Selection Shaped How Humans Reason? Studies with the Wason Selection Task*, "Cognition" 1989, vol. 31, no. 3, pp. 187–276. As Johnson-Laird notes, if the problem concerns a "sensible everyday generalization", people are also much less prone to errors; see P.C. Wason, P.N. Johnson-Laird, *Psychology of Reasoning: Structure and Content*, Harvard Univ Pr, 1972; P.N. Johnson-Laird, *Deductive Reasoning*, "Annual Review of Psychology" 1999, vol. 50, pp. 109–35.

of the Wason task are real, the traditional interpretation of the phenomenon might be misguided.[49]

In what follows, I will assume that the explanation of the Wason task phenomenon should explain not that the answer offered by most subjects is incorrect (which would involve the controversial assumption that the material implication is the norm) but that human reasoning capacities are context-sensitive as shown by the evidence. From this perspective, the main point of the NEF model is to show that it is able to describe compositional phenomena such as inference and its context-dependence in a biologically-realistic way. In other words, the model answers the challenge of Fodor and Pylyshyn who claimed that neural networks cannot properly deal with systematicity or productivity of human language.[50] In some ways, answering this theoretical challenge might be considered more pressing a need than modelling a real phenomenon whose exact nature is still debatable.

The theoretical challenge is also the reason why I analyze it: I want to inquire whether the explanation of the phenomenon is general enough to tackle the problem of structured representations. Fodor and Pylyshyn claimed that connectionism must face a dilemma. Either connectionist representations lack the structure needed to describe logical relationships that underlie compositionality of mental representations, or they should be considered mere implementations of high-level symbol structures. The second horn of the dilemma can

[49] K. Stenning, M. Van Lambalgen, *Human Reasoning and Cognitive Science*, MIT Press, Cambridge, Mass. 2008. In particular, there seem to be many different cognitive phenomena that call for explanation, and the evidence for them is lost when the results of individual participants are averaged. Also, it is interpretation of the linguistic form that is context-dependent. Note that the NEF explanation might be understood to deal with context-dependence of the interpretation of the linguistic form rather than of the reasoning. However, since there are multiple phenomena involved in different ways participants solve the task, the model described is not really explanatory for all of them. Note that it is not really essential for my argument to claim that BioSLIE really explains all these phenomena. Its interest for me is mostly in the way the NEF tackles structured representations and inferences.

[50] J.A. Fodor, Z.W. Pylyshyn, *Connectionism and Cognitive Architecture: A Critical Analysis*, "Cognition" 1988, vol. 28, no. 1–2, pp. 3–71.

be rephrased in terms of unification: the high-level symbol theory will unify the phenomena better than an implementation-level theory, so the latter will be theoretically disadvantageous. A proper answer to the challenge has to question the assumptions made by Fodor and Pylyshyn; otherwise, it won't be satisfactory. Hence, it should be shown that, on the one hand, it is possible to explain such inference-related phenomena like context-dependent reasoning on the neural level, and on the other hand, that such explanations are not overridden by higher-level symbolic models. If the NEF model has a satisfactory answer, we might be able to conclude that the level of unification is sufficient for offering a level of generality required of cognitive theories, and that at the same time it offers sufficient detail to be informative of context-dependent inference processes.

A model of the human reasoning capabilities underlying the results made so salient by the Wason selection task was dubbed BioSLIE (for BIOlogically-plausible Structure-sensitive Learning Inference Engine). The critical part of the model is the choice of representational format. To this task, Holographic Reduced Representations (HRRs) were chosen.[51] 100-dimensional HRR vectors are used to encode linguistic structure by using circular convolution (⊗); decoders use circular correlation (⊕). Again, the exact mathematical detail need not interest us here. Suffice it to say that it is relatively straightforward a matter to use the mathematical definitions of HRRs to derive a specific kind of vector representations in the NEF models using LIF neurons. The HRRs are then recruited to encode the inferences in the manner typical for production systems. In general, an implication "if p then q" will be encoded in the following way:

$$R = \text{relation} \otimes \text{implies} + \text{antecedent} \otimes p + \text{consequent} \otimes q$$

The rule "If there is a vowel on one side of the card, there is an even number on the other side" can be written as:

[51] For background, see T.A. Plate, *Holographic Reduced Representation: Distributed Representation for Cognitive Structures*, CSLI Publications, Stanford, CA 2003.

R = relation ⊗ implies + antecendent ⊗ vowel + consequent ⊗ even

Now the rule has to be computed, or transformed, to replicate the experimental result, using the following transformation:

T1 = antecedent' + implication' ⊗ relation' + consequent'

since T1 ⊗ R ≈ vowel + even. Eliasmith notes:

> It is here, in constructing our representation R in this manner, that we avoid merely implementing a symbolicist system. This is because this representational format, being a compressed vector representation, does not explicitly include the constituents of the representation R in the representation itself. As a result, the representation is non-compositional, violating a basic constraint Fodor and Pylyshyn (1988) place on symbolicist cognitive systems.[52]

In other words, Eliasmith escapes the dilemma posed by Fodor and Pylyshyn by proposing a third way: HRRs only approximate a logical structure of propositional inference but because of the compression of information involved, the decoding process can only restore part of the original structure. For this reason, compositionality is only approximately implemented; still, Fodor and Pylyshyn might say that they are free to idealize these neural details away.

It is known that the subjects choose the answer proper for the material conditional in the context where permission is involved (like the permission to drink alcoholic drinks, as in the above example). For that to be modelled, some mechanism of context-switching must be used. As BioSLIE is biologically realistic, it will refer to known functional anatomical facts about possible sources of switching. The network in the model consists of three parts:

---

[52] C. Eliasmith, *Dynamics, Control, and Cognition*, [in:] *Cambridge Handbook of Situated Cognition*, eds. P. Robbins, M. Aydede, Cambridge University Press, Cambridge 2009.

(1) input from ventromedial prefrontal cortex (VMPFC) which provides familiarity, or context, information that is used to select the appropriate transformation;
(2) left language areas which provide representations of the rule to be examined;
(3) anterior cingulate cortex (ACC) which gives an error signal consisting of either the correct answer, or an indication that the response was correct or not.[53]

The neural populations modelled by BioSLIE are parts of the right inferior frontal cortex, where VMPFC and linguistic information is combined to select and apply the appropriate computation to solve the Wason selection task. Thanks to this structure, the model is able to appropriately switch transformation rules depending on the context: in the social context, the rule proper for material implication is used (if p then q → {~q, p}), and in the abstract context, an 'incorrect' rule (if p then q → {q, p}). For this reason, it is able to reproduce the typical results from the Wason task under both contexts. Notably, the system does not simply store a solution (like a hash table) but generalizes: it can transform novel structured representations based only on their syntax.

At the same time, there are some crucial idealizations in the model. For example, there is no detailed explanation of how linguistic form of representations is used to switch the context (though it stands to reason that it can be done purely syntactically, given the successful performance of lowly syntax-level 'sentiment analyzers' such as The General Inquirer[54]) or why we have different contexts to start with. It also models human performance largely at the behavioural level, while the neurological level remains grossly simplified. For example, there is no easy way to include the evidence about response

---

[53] See *ibid.* for references from literature that justify these choices.

[54] P.J. Stone et al., *The General Inquirer: A Computer System for Content Analysis and Retrieval Based on the Sentence as a Unit of Information*, "Behavioral Science" 1962, vol. 7, no. 4, pp. 484–498. Note however that it still remains to be explained why certain words get classified this way, and not another.

time, which is one of the important factors in deciding whether a task was easy for the subject. The model does not represent the human brain to such a detail: BioSLIE seems to perform much faster than a biological neural network and this is not because of the faster electronic hardware (hardware speed, in general, is a negligible factor if the complexity is the same[55]). For this reason, it is closer to symbolic models than to dynamic ones.

The model seems to fit the previous pattern quite neatly; we could also create a second argument pattern to compare a traditional symbolic solution – for example, a domain-general simulation of a reasoning theory (the one that Fodor and Pylyshyn would approve) or a context-sensitive modular reasoning theory in the style of evolutionary psychology – and BioSLIE. Unfortunately, this is not entirely possible just because the ideas of evolutionary psychologists are not implemented. There are several domain-general solutions, such as relevance theory[56] or analysis based on information-gain,[57] but the space prohibits any detailed comparisons here. The most important difference between symbolic models, on the one hand, and BioSLIE is that the latter one does not presuppose that reasoning is actually realized by an algorithm as powerful as classical logic (or at least close enough). In other words, instead of accepting Fodor's challenge, it stresses that compositionality and systematicity might be unsubstantiated theoretical posits. In particular, even if for some purposes it might be useful to idealize human reasoning competence as classical logic (something that many logicians eschew today[58]), our performance is better explained as only roughly approximating it. The

---

[55] See for example I. Van Rooij, *The Tractable Cognition Thesis*, "Cognitive Science" 2008, vol. 32, no. 6, pp. 939–84.

[56] D. Sperber, F. Cara, V. Girotto, *Relevance Theory Explains the Selection Task*, "Cognition" 1995, vol. 57, no. 1, pp. 31–95.

[57] M. Oaksford, N. Chater, *A Rational Analysis of the Selection Task as Optimal Data Selection*, "Psychological Review" 1994, vol. 101, no. 4, pp. 608–631; M. Oaksford, N. Chater, *Information Gain Explains Relevance which Explains the Selection Task*, "Cognition" 1995, vol. 57, no. 1, pp. 97–108.

[58] K. Stenning, M. Lambalgen, *Human Reasoning and Cognitive Science*, *op. cit.*

actual process of inference is rather noisy, and a trained logician might solve the original abstract task in a way that conforms to the interpretation of the rule as instantiating a material implication, so the context-switching cannot be too strict. Similarly, the model might solve some of the tasks 'correctly' because of the noisy decoding and approximation.

It may be therefore observed that BioSLIE is not a mere implementation of a symbolic system. At the same time, it seems that even if the NEF methodology does not include particular information – there is no unifying theory of memory and decision-making in the background, in contrast to Rolls' conception – it is able to include data from various sources. Rather than being shallow by design, the theory seems to require researchers to include important available evidence in the models, such as the evidence about what might be the source of familiarity information. For this reason, even if the NEF is not a detailed theory about a human brain, it does not seem to be general just because of the low information content.

Let me sum up this part of the chapter. I have studied two different examples – explanations of different phenomena – implemented as artificial neural networks according to the NEF principles. The principles make the process of implementation relatively systematic, even if they are silent about the particular functional structures in real biological brains. This silence has its source, of course, in the fact that it is still very hard to generalize facts about architectures of brains to the extent that would make it possible to create a grand unifying theory. But it may be already said that the NEF satisfies the requirements of Kitcher's theory of explanation, although its explanatory power cannot be explicated solely in terms of unification.[59]

---

[59] Notably, in his forthcoming book *How to Build a Brain* (Oxford University Press), Eliasmith develops tools that address the abovementioned shortcomings of the NEF: for example, they allow the inclusion of the response time data in the model. As for higher cognition, it is hypothesized now that it is made possible by semantic pointers, which are neural representations that carry partial semantic content and are composable into the representational structures.

Conclusion

I have reviewed the unification theory of explanation, and applied it to analyze two biologically realistic models of cognitive processing, which were built according to the Neural Engineering Framework. I noted that the theory of explanation as such still faces important problems but it can be, at the very least, used to define the feature that many good theories in science share, namely having a large scope, or breadth.

Could the unification theory somehow be improved so as to be useful in such cases as overabundance of competitive explanations, as in the case of the Wason selection task? Indeed, it seems that the methodology of psychology needs to rely on high-level criteria of selection of explanatory models, and some such criteria have already been proposed.[60] As can be observed, most of the criteria proposed by statisticians or practitioners in modelling do not refer to argument patterns, or anything similar, for that matter. Even if argument patterns seem familiar for most philosophers – logic is one of the main tools in philosophy, after all – it does not seem to be a particularly good idea to require that all theories need to be subsumed under a language-like format to compare them. For a start, computational models might be really hard to translate in detail to a scheme expressed in predicate logic. The sketch of the argument pattern I offered above is probably intuitive but is it really a faithful representation of the NEF modelling? Surely, if we have quantitative models, we should be able to compare them precisely, given how specific their predictions are, instead of relying on rough verbal approximations.

I have already suggested that we might refer to the notion of information, and methodology already makes use of Kolmogorov complexity criteria (and other related measures of complexity,

[60] M.A. Pitt, I.J. Myung, S. Zhang, *Toward a Method of Selecting Among Computational Models of Cognition*, "Psychological Review" 2002, vol. 109, no. 3, pp. 472–491; J.R. Busemeyer, A. Diederich, *Cognitive Modeling*, Sage, Los Angeles 2010.

such as Aikake Information Criterion or Bayesian information criterion). Without going into the detail of statistics or complexity theory, we can say that they allow the operationalizing of the requirement that a theory has both sufficient breadth and depth. Indeed, without specifying that a theory is not shallow, the account of explanation in terms of unification might be trivially satisfied by a theory devoid of any empirical content, or with just minimum content. By using the formalism of logic, of course, proponents of the CL account can easily exclude, as they do, logical truths from the premises of explanatory arguments. What they cannot do is to exclude trivial but non-logical truths, such as the one that everything is located in space. Granted, these truths are useless to explain many phenomena, but are really general, so they should be treated as generalizations that have some explanatory power. It is obvious however that they are shallow; and explaining why that is the case seems to be feasible in information-theoretic terms. But this is a topic for another occasion.

Let me return to the main topic. It seems that the NEF is one of the few theories in neuroscience that answers the need for a general frameworks and conceptions. While it does not integrate all kinds of empirical data that we already have, it seems to be, indeed, a general and sufficiently deep framework to deserve to be called a theory. It can be used to model both higher-level cognitive tasks, such as inference, and low-level processes that underlie sensomotoric behaviours, such as path integration. Even if the model of inference was highly idealized, this idealization is relaxed in the currently developed extensions of the NEF, which shows that it can address realistically even a wider scope of cognitive and neural phenomena. Of course, this alone does not make it the only candidate theory for unification in neuroscience; such a conclusion would definitely be premature. All too often, we are still in the dark about how the brain works; yet it does not mean that there is nothing that we can learn from neuroscience or that we should stop attempting to unify our extant knowledge.

Radosław Zyzik
Jesuit University of Philosophy and Education *Ignatianum*
Copernicus Center for Interdisciplinary Studies

Biological Determinism and Its Enemies[1]

Introduction

Biological determinism implies that all human behaviour is programmed and controlled by a chain of biological factors. Human actions are fixed by the biochemical properties of cells that build every single person.[2] Biological determinism has generated much attention over the last three decades. Due to the development of neuroscience, genetics and many other biological sciences, scientists are able to identify the biological basis of a large amount of human behaviour. For many years biological determinism was only a philosophical hypothesis without any empirical basis. Today, as we learn more about the role of our genes and nervous systems, part of the scientific community is inclined to assert that the deterministic view of human nature is the correct one. A striking feature of this philosophical hypothesis are its possible political and moral implications.

Biological determinism can serve as an excuse for social inequalities, because social inequalities are the consequences of differences in abilities between individuals. Societies are just a collection of individuals, and individuals in turn are collections of biological

---

[1] This paper was written within the research grant "The Limits of Scientific Explanation" sponsored by the John Templeton Foundation.

[2] S. Rose, R. Lewontin, L.J. Kamin, *Not In Our Genes: Biology, Ideology and Human Nature*, Penguin Books Ltd., Harmondsworth 1990, p. 5.

molecules. Therefore, the structure of society is natural and it should not be changed.[3] Biological determinism also poses a threat to our moral convictions. Without going into details, we can say that people are held responsible for their actions because they are moral agents. Agents are capable of reflecting on their situation, forming intentions about how they should act and they are able to act based on reasons. According to biological determinism, people are not moral agents, because they are compelled by biological factors, i.e. genes or brains.[4]

Little research (if any) has addressed the problem of determinism from more than one perspective at the same time. On the one hand, one can read about the neuroscience of free will and the renaissance of determinism due to the work of neuroscientists. On the other, a new face of genetic determinism is discussed as a result of the progress made in genetics. Moreover, today we can also learn about the impact of biological factors on the development of model organisms in neurogenetics. With this in mind, we have tried to investigate how determinism is understood in neuroscience, behavioural genetics and in a new discipline which combines knowledge from many disciplines – neurogenetics.

We believe that only such a broad perspective will eventually allow an understanding of determinism in biology with all of its shortcomings. Therefore, the aim of our study is to evaluate the philosophical interpretations of neuroscientific, genetic and neurogenetic experiments that can be seen to be in line with the thesis of biological determinism. The paper re-examines the tacit philosophical assumptions, applied methodology and interpretation of the results of the experiments.

In the first section of the paper we focus on the neuroscientific experiments whose interpretations may endorse the deterministic thesis.

---

[3] R. Lewontin, *Biological Determinism*, "Tanner Lectures on Human Values" 1983, vol. 4, pp. 155–156.

[4] Cf. Ł. Kurek, *Problem wolnej woli z perspektywy nauk empirycznych*, "Logos i Ethos" 2011, vol. 30, pp. 99–138.

We analyze Benjamin Libet's famous experiments and its further redesigns. In the second section of the paper, we investigate the PKU story, which in behavioural genetics is a paradigm case for anyone interested in identifying the effects of genes on human behavioural traits. In the third part of the paper, we present the most complex view on the role of genes, nervous systems and environment on the behavioural traits. This view can be found in neurogenetics, especially with regards to the studies on model organisms.

We believe that the most fearsome enemies of biological determinism can be found not in philosophy, but in biology. Our knowledge about the role of biological factors in human behaviours is fragmentary, incomplete and full of simplifications and speculations. Therefore we would like to present three paradigm cases: Libet's experiment, the PKU story and *C. elegans* case in order to show that even in these cases there is no reason to consider biological factors as puppet masters pulling our strings.

1. Neuroscience in the Service of Determinism?

To date neuroscientists have conducted a large number of experiments, trying to establish the role of biological factors in human decision making. The most well-known, influential and controversial experiment conducted by neuroscientist is the one designed by Benjamin Libet, the American psychologist, who claims that the real causes of our decisions are in fact the subconscious processes in our brains.

We have chosen this experiment and its later refinements as subjects of analysis for several reasons. Firstly, Libet's experiment was the first that tried to investigate the relation between conscious and subconscious processes in human decision making.[5] Secondly, the experiment has been redesigned many times after. Thirdly, the impact of

[5] B. Libet, *Do We Have Free Will?*, [in:] *Conscious Will and Responsibility*, eds. W. Sinnott-Armstrong, L. Nadel, Oxford University Press, Oxford 2011, pp. 1–11.

Libet's work goes far beyond neuroscience. The different aspects of it are discussed by philosophers, ethicists and even lawyers.[6] Fourthly, the impact of the experiment is not limited merely to the amendments to a few philosophical theories. On the contrary, some philosophers and scientists claim that Libet has provided an ultimate proof in favour of biological determinism and, simultaneously, against all understandings of free will.

In this section of the paper we would like focus on presenting and commenting on the methodological and conceptual aspects of Libet's experiment and its redesigns, as well as the philosophical assumptions made by Libet and his critics. We will start our investigations by presenting Benjamin Libet's experiment and his interpretation of its result. Then we refer to the most popular and insightful critical remarks presented both by scientists and philosophers. In the next part we will briefly present redesigns of Libet's work made by Miller and Trevena,[7] Banks and Isham,[8] and one of the most up to date experiments conducted by John-Dylan Haynes and his colleagues.[9]

Before we move on, we would like to acknowledge one distinctive feature of every discussion related to the issue of biological determinism in neuroscience. Discussing this issue from a neuroscientific perspective is slightly different than discussing it in neurogenetics or in behavioural genetics. The main difference lies in the conceptual scheme. While in genetics and neurogenetics we debate the role of biological factors in our life using a concept of biological determinism from time to time, in neuroscience the concept of free will is at the

---

[6] Cf. L. Alexander, *Criminal and Moral Responsibility and the Libet Experiments*, [in:] *Conscious Will and Responsibility*, eds. W. Sinnott-Armstrong, L. Nadel, Oxford University Press, Oxford 2011, pp. 204–207.

[7] Cf. J.A. Trevena, J.G. Miller, *Cortical Movement Preparation Before and After a Conscious Decision to* Move, "Consciousness and Cognition" 2002, vol. 10, pp. 162–90.

[8] W.P. Banks, E.A. Isham, *We Infer Rather Than Perceive The Moment We Decided to Act,* "Psychological Science" 2009, vol. 20, pp. 17–21.

[9] C.S. Soon, M. Brass, H.-J. Heinze, J.-D. Haynes, *Unconscious Determinants of Free Decisions in the Human Brain*, "Nature Neuroscience" 2008, vol. 11, pp. 543–545.

heart of the debate.[10] This feature of discussions is very easy to explain. It is not really a surprise that neuroscientists interpret their findings in terms of conscious intent, subconscious processes, volition and – what comes straightforwardly – the concept of free will. While their work is focused on the brain systems associated with the decision making process, an interpretation of their results is strictly philosophical.[11] Philosophy was interested in the nature of human activity long before biology, not excluding the ability to making a free decision. Therefore, philosophers delivered conceptual schemes to scientists which were picked up by them and used to describe results of their works.

However, we strongly believe that there is no need to use this highly controversial and vague philosophical concept while discussing the role of our brains. One of the reasons for this conceptual prohibition is a lack of a precise definition of 'free will'. In philosophy, one can encounter a different understanding of the concept, including concepts which are contradictory to each other. This is one of the reasons that there is no single experiment in neuroscience that can support or suppress all understandings of free will. It means that scientists may design and conduct their experiments for the testing existence of the one and only understanding of 'free will'. And yet it seems that they have no problems with claiming that all forms of free will proved to be incorrect (or supported). We have to admit that the state of art forces us to use the concept of free will in a few places, although we will use it as rarely as possible.

In order to avoid the constant need to define the concept of free will we will use the 'biological determinism' notion which have been defined at the beginning of the paper. In the case of neuroscience, the biological factors that are in control of our selves are our brains. Let's see now if the results of the experiments conducted by Libet and his followers really indicates what their authors claim they are.

---

[10] K. Smith, *Neuroscience vs Philosophy: Taking Aim at Free Will*, "Nature" 2011, vol. 477, pp. 23–25.

[11] Cf. M.R. Bennett, P.M. Hacker, *Philosophical Foundations of Neuroscience*, Wiley-Blackwell, Oxford 2003.

1.1. Benjamin Libet and the Beginning of the End of Free Will?

The first and most influential experiment was conducted by Benjamin Libet in the 1980s.[12] Libet has set himself a goal to determine the role of conscious and subconscious processes in making simple motor movements. His work on the neural basis of simple decisions and their relation to conscious feeling has set the stage for further research devoted to these problems.

Libet used an electroencephalogram (EEG) to measure changes in the electrical activity of brains. In his experiment he put electrodes on the scalps of subjects to measure the electrical activity of the cortex. The higher voltage level noted by electrodes is supposed to suggest that particular neuronal systems in the cortex are more active. Participants were asked to flick their wrists or to move their fingers and they were informed that this was their task and they had to choose random moments to move or flick. The moment of this simple motor movement was measured with an electromyography (EMG), a device used to record the muscle action with electrodes placed on skin. In order to identify a moment of conscious decision making, subjects were asked to remember a position of the dot on the oscilloscope timer when they started to be aware of the desire to move a wrist (or finger) and also were asked to press the button to note the position electronically.

It is a well-known fact that the level of brain's electrical activity rises before an action, and this preceding activity has been called the readiness potential.[13]

Libet asked himself a question:

[12] B. Libet, C.A. Gleason, E.W. Wright, D.K. Pearl, *Time of Conscious Intention to Act in Relation to Onset of Cerebral Activity (Readiness-Potential). The Unconscious Initiation of a Freely Voluntary Act*, "Brain" 1983, vol. 106, pp. 623–642.

[13] Cf. H.H. Kornhuber, L. Deecke, *Hirnpotentialänderungen bei Willkürbewegungen und passiven Bewegungen des Menschen: Bereitschaftspotential und reafferente Potentiale*, "Pflügers Archiv European Journal of Physiology" 1965, vol. 284, pp. 1–17.

> (...) *when* does the *conscious* wish or intention (to perform the act) appear? In the traditional view of conscious will and free will, one would except conscious will to appear before, or at the onset of, the RP, and thus command the brain to perform intended act.[14]

Libet's goal was to compare the moment when the readiness potential appears with the felt intention to move. He discovered that the readiness potential is present almost 350–400 milliseconds *before* the moment when subjects reported the felt intention to move. Moreover, the further 200 milliseconds passed *by the time* the actual move was made. Therefore, the 550–600 milliseconds which separates the moment when the readiness potential appears and the moment of actual action. Libet's own interpretation of his findings was this – people do not have free will, because the decision to act is made by the brain and is beyond conscious control.

The timing issue is really at the heart of the Libet's argument. A decision to move, that was consciously felt to be free, appeared to be predetermined by subconscious brain processes. Therefore, the conscious brain processes are not the real causes of our decisions. The real causes are the subconscious processes in our brain, which precede any conscious activity associated with an action. However, Libet asked himself if there is any role for conscious will in the performance of a voluntary act? He introduced the concept of the "free veto" – though the conscious will is not able to initiate a decision, it is able to control whether such a decision can take place:

> Potentially available to the conscious function is the possibility of stopping or vetoing the final progress of the volitional process, so that no actual muscle action ensues. *Conscious-will could thus affect the outcome* of the volitional process even though the latter was initiated by unconscious cerebral processes. Conscious-will might block or veto the process, so that no act occurs.[15]

[14] B. Libet, *Do We Have Free Will?*, [in:] *Conscious…*, *op. cit.*, p. 2.
[15] *Ibid.*, p. 5.

Our brains are the ultimate causes of our behaviours, but *We* are able to suppress or mediate any given act or decision. There is a problem here. People have less than 200 milliseconds to change their action. This is a time window between being aware about decision and performing an action. According to Libet, we have a "free veto", but a very short time window in which to use it. If we miss it, then our brains have really made the decision in question. This unintuitive conclusion is a result of accepting a particular understanding of "free will":

> Some have proposed that even an unconscious initiation of a veto choice would nevertheless be a genuine choice made by the individual and could still be viewed as a free will process. I find such a proposed view of free will to be unacceptable. In such a view, the individual would not consciously control his actions; he would only become aware of an unconsciously initiated choice. He would have no direct conscious control over the nature of any preceding unconscious processes.[16]

To sum up, Benjamin Libet was the first scientist who took an experimental perspective investigating the neuronal basis of human decision making and its relation to the executions of conscious will. Credit must be given to Libet for paying attention to the neuronal foundations of our decision making process. Many other experiments have followed the initial thoughts of Libet that our subconscious processes are the real puppet-masters pulling our strings.

---

[16] *Ibid.*, p. 5.

1.2. 7 Seconds Ahead. The Final Blow for Biological Determinism?

John-Dylan Haynes, when he was designing his experiment, was aware of the shortcomings of Libet's work.[17] In order to avoid at least few of them he decided to investigate the relation of RP with other brain regions. Especially, he was convinced that prefrontal brain regions can be better predictions for motor movements than the brain regions identified by Libet.[18]

In his experiment Haynes used different tools to measure brain activity and the decision making moment. To identify brain activity he used functional magnetic resonance (fMRI) and not EEG as in Libet's case. Moreover, he replaced the clock with a randomized stream of letters that changed every 500 milliseconds. The subject had to report the letter that was on the screen at the moment of making a conscious decision. Subjects were asked to push the left or right button while lying in the MRI scanner.

In the next two steps, different brain regions were identified. In the first step, they identified brain regions that had information about a subject's decision after it had been made. Secondly, they recognized which brain regions had predictive information about a subject's decision, even before the subject knew how they were going to decide. Two brain regions was found to be responsible – *frontopolar cortex and precuneus/posterior cingulate cortex*, which had predictive information already 7 seconds before the decision was made:

> The conscious decision to push the button was made about a second before the actual act, but the team discovered that a pattern of brain activity seemed to predict that decision by as many as seven seconds. Long before the subjects were even aware of making a choice, it seems, their brains had already decided.[19]

[17] J.D. Haynes, *Beyond Libet: Long-Term Prediction of Free Choices from Neuroimaging Signals,* [in:] *Conscious…*, *op. cit.*, pp. 86–87.

[18] *Ibid.*, p. 87.

[19] K. Smith, *Neuroscience…*, *op. cit.*, pp. 23–25.

It could mean – according to one possible manner of interpretation – that the high-level networks in the brain are active long before we made a conscious decision. The result surprised Haynes himself:

> <The first thought we had was ‘we have to check if this is real>, says Haynes. <We came up with more sanity checks than I’ve ever seen in any other study before>.[20]

However, Haynes and his colleagues did not find ultimate proof for the determinacy of free choices. His prediction is statistically reliable but it is far from perfect. They were able to predict the decisions of subjects only with an accuracy of 60%. Moreover, Haynes and his colleagues made tacit philosophical and methodological assumptions which should be addressed before any conclusive fact about free choices can be proposed.

Haynes followed in Benjamin Libet’s footsteps. He has a similar understanding of free will and made similar assumptions about the human ability to report the moment of conscious intention to act and many more. Libet’s work are really the same in its nature. It is true that Haynes used more sophisticated tools to identify the neuronal basis of human decision making, but his goal and his philosophical and methodological assumptions remain the same as in the 1980s when Libet’s paper was published. One should not look for a new version of an old experiment, but try to recognize the tacit assumptions made by Libet and then critically analyze them in order to answer the question of whether our brains are the actual causes of our behaviour.

1.3. The Triumph of Biological Determinism. Oh, Really?

Libet and his followers searched for a neuronal basis for human decision making. It must be said that their work founded and constituted a new subdiscipline of neuroscience, namely the neuroscience

[20] *Ibid.*, p. 25.

of free will. Nevertheless, there are several reasons to be skeptical about their work. In the next part of the paper we will focus on Libet's tacit assumptions and philosophical presuppositions. We will conduct our analysis from two interconnected perspectives: a philosophical and an experimental one. Referring to the work of such notable thinkers as Daniel Dennett, Adina Roskies or Al Mele, we are going to focus on the conceptual and methodological aspects of the experiments. On the other hand, taking an experimental stand, we will discuss two issues that can be found in arguments delivered by both Libet and Haynes. Namely the role of RP and the subject's ability to perceive and report the moment of the appearance of conscious will. The first experiment focuses on the famous readiness potential and its role in decision making. Miller and Trevena propose another explanation for the appearance of the RP, where RP is not really the cause of an action, but a signal of the brain paying attention. The second experiment conducted by Banks and Isham is devoted to the subject's ability to report the moment of becoming aware of an intention to act. They show that the moment is not *perceived* but *reconstructed* from further information input.

The scientific community has been divided since Libet's paper was published. Some philosophers uncritically accepted Libet's findings and announced that people are not free agents.[21] We should start rethinking the fundamentals of our moral, ethical and legal systems. Our whole conceptual scheme in the normative sciences, and in religion, has to be changed, because they have been based upon the concept of free will, and Libet's experiment has proven that no such thing exists, therefore we have to rethink our belief systems.

On the other hand, a large group of philosophers remained skeptical and even suspicious with regards to Libet's findings. One of the most prominent critic of Libet was Daniel Dennett, who focused on both conceptual and methodological issues. Dennett agreed that readiness potential (RP) was measured objectively. But the moment of the

[21] Cf. D. Dennett, *Freedom Evolves*, Viking London 2003, pp. 221–257.

appearance of the conscious desire to act was measured based on the subjective reports of the participants. Participants, according to Dennett, did not note the actual moment of the conscious decision, but only what they *seemed* to be conscious of the desire to act. There is no obvious reason for assuming that when someone is *feeling* that she is making a decision that it is objectively the same moment as the taking of a conscious decision. Moreover, the subjects had to shift their intention from the their inner feelings of the will to the oscilloscope timer. This can result in time mismatches. According to Dennett, the experiment is inadequate when it comes to establishing time relations between RP and the conscious intention to act, because the appearance of a conscious intention to act was measured improperly.[22]

This was a methodological objection, but Dennett also takes a closer look at the philosophical presuppositions that stand behind Libet's own interpretation. Dennett seems to suggest that the philosophy involved in the experiments is an example of bad thinking.[23] He tries to reconstruct Libet's philosophical presuppositions and localize the *You* in the brain, because such a presupposition seems to be accepted by Libet. He proposes three places where *You* could have its localization, but at the end, none of these places is sufficient to fulfil Libet's understanding of *You*. There is no such *You* in the brain and

[22] D. Dennett, K. Marcel, *Time and the Observer*, "Behavioral and Brain Sciences" 1992, vol. 15, pp. 183–247.

[23] "Since Libet wants to hear from you, not your striate cortex, we have to know where you are in the brain before we can even begin to interpret the data. Let us suppose, for the sake of argument, that this makes sense. To be fair and constructive, cast aside all the extravagant versions of the supposition: Libet is not supposing that you are an actual homunculus, with arms and legs, eyes and ears, like the little green man in the control room of the man-size puppet in the morgue in *Men in Black*, and he's not supposing that you are an immaterial portion of glowing ectoplasm that oozes around in your brain like a ghost amoeba, or that you are an angel whose wings are folded till you are called to fly to heaven. We must consider a minimalist version of the hypothesis, stripped of all-such embarrassing details: You are just whatever-it-takes-to-be-able-to-experience--decision-and-clock face-orientation-simultaneity. (If we need to have an image, we can dimly imagine that this whatever-it-is is some nexus or cluster of brain activity, and it might shift around under various conditions, a brainstorm with rather special cognitive powers)", D. Dennett, *Freedom Evolves…*, *op. cit.*, pp. 232.

this is the effect of not only a bad philosophy, but perhapse also a misunderstanding of brain functioning.[24]

Michael Gazzaniga, one of the most known and prominent neuroscientist, claims that such studies as Libets and Haynes are based upon assumptions about brain activity that could be incorrect. These studies suggest that scientists presuppose that our brains function in a linear way. Step after step after step is taken in order to execute a decision or an action. One step is a result of the preceding step and so on. Gazzaniga suggests that the brain is a set of processes working in parallel, complex networks that communicate between each other constantly. Such a vision of the brain leads him to the conclusion that the time at which one is aware of a decision is not so important. As Dennett writes, there is no such place as *You* where everything begins:

> Once you distribute the work done by the homunculus (in this case, decision-making, clock-watching, and decision-simultaneity-judging) in both space and time in the brain, you have to distribute the moral agency around as well. You are not out of the loop; you are the loop. You are that large. You are not an extensionless point. What you do and what you are incorporates all these things that happen and is not something separate from them. Once you can see yourself from that perspective, you can dismiss the heretofore compelling concept of a mental activity that is unconsciously begun and then only later "enters consciousness" (where you are eagerly waiting to get access to it). This is an illusion since many of the reactions you have to that mental activity are initiated at the earlier time—your "hands" reach that far, in time and space.[25]

It is not only Daniel Dennett who remains skeptical about such neuroscientific experiments. Adina Roskies does not share Libet's belief that spontaneous motor movements can be taken as a paradigm for

---

[24] *Ibid.*, p. 133.

[25] D. Dennett, *Freedom Evolves*..., *op. cit*, p. 242.

free choices. In her enlightening paper she claims that the burden of proof is on Libet and his followers, because they refer to motor action as a paradigm type of free decision and they have drawn the conclusion that their findings can easily carry over to every other kind of decision, even to the most complex ones.[26] She claims that such arbitrary actions are, in the best case, a degenerate example of free will at work, because what really matters are the decisions made for particular reasons. We are free agents capable of acting for reasons; in the experiments, the subjects were not acting for reasons at all. Therefore, the authors of such studies should present reliable arguments in favour of spontaneous motor movements, because only then we will be able to see them as a paradigm case for every decision. Until that time, drawing such a general conclusion, i.e. there is no free will, only a free veto, is far too premature.

Alfred Mele focused on the conceptual part of the experiments in his work and he noticed the ambiguous nature of the concepts used in the interpretation of the works. This may appear to some as an example of splitting hairs, but it is not. In scientific research when one has to investigate something, one has to define carefully what they are really pursuing. In Libet's experiment, such conceptual analysis has not been made. Therefore the subjects were left on their own without defining the awareness of the conscious will. How one can measure something when one is not sure what it means?[27] Moreover, what is the difference between "wanting to do something", "indenting" and at last, "deciding" to act? He cites Libet's paper when he occasionally asks the subject to recall the clock position when they feel the awareness of a decision, intention, urge, wanting, will or the wish to move.[28]

---

[26] A. Roskies, *Why Libet's Studies Don't Pose a Threat to Free Will*, [in:] *Conscious…*, *op. cit.*, p. 17.

[27] Cf. A. Mele, *Libet on Free Will: Readiness Potentials, Decisions, and Awarness*, [in:] *Conscious…*, *op. cit.*, p. 29.

[28] Cf. A. Mele, *Springs of Action: Understanding Intentional Behavior*, Oxford University Press, New York 1992; A. Mele, *Motivation and Agency*, Oxford University Press, Oxford 2003.

Dennett, Roskies and Mele are philosophers and their criticism has a philosophical flavour. They have criticized the conceptual and methodological aspects of the experiment from a philosophical perspective. However, because Libet's work is still highly popular amongst supporters of biological determinism in science and philosophy, few experiments have been conducted to further investigate the role of RP and the awareness of the intent to act in decision making process.

Jeff Miller and Judy Trevena in 2009 redesigned Libet's experiment. They also used scalp electrodes (EEG) to measure the readiness potential of the subjects. However, when Libet let the subject decide when to move their wrist, they informed their subject to wait for an audio tone *before* deciding to tap a key or not. Following the interpretation of Libet's experiment, the RP should be greater when a person chose to push the key. RP should be the main cause of the action. Researchers noted that recorded celebral activity (RP) was at the same level regardless of the course of the decision made by subjects. There was no difference between subjects who chose to tap the key after hearing the tone and the subjects that chose not to tap the key.

The interpretation of the experiment proposed by Miller and Trevena states that RP is just an electrical sign that brain is paying an attention:

> If movement-preceding negativity (RP in Libet experiment – R. Z.) reflects the brain's preparation to move – conscious or otherwise – then people should be more likely to move when this negativity is large than when it is small. In two experiments, we found no evidence of this predicted difference, despite observing clear negativity. We conclude, then, that such negativity does not necessarily reflect preparation for movement, and that it may instead simply develop as a consequence of some ongoing attention to or involvement with a task requiring occasional spontaneous movements. For example, although the precise moment of tone onset was unpredictable (because of the exponential distribution of onset times) it was predictable that the tone would probably occur

> sometime during the trial interval, and the negativity could represent a temporally-sustained preparation for an expected onset at some unspecified time during the trial.[29]

Therefore, RP (or movement-preceding negativity) was mistakenly interpreted as a process responsible for decision making. RP precedes every decision, but it does not mean that the assumption that RP is the real cause of our action is the correct one. Especially, it does not have anything to do with the *content* of any given decision. It is just a signal that the brain is paying an attention.

Miller and Trevena conducted a second experiment to check Libet's assumptions that subconscious processes determine the content of our decisions. They asked subjects to press the key after a tone and to decide on the spot which hand they would use to tap it. The movement of the right hand is associated with left hemisphere activity and the movement of a left hand is associated with the right hemisphere. They hadn't noted the correlation in question and were not able to identify, based on the brain activity, the hand which the subject would use to press the key, *before* they actually pressed it:

> This conclusion is clearly at odds with the fact that such negativity is always a reflection of the brain's subconscious preparation for movement. Instead, our results converge nicely with several previous findings casting doubt on such negativity as a direct measure of movement preparation, and thus further weaken a crucial link in the argument that Libet results prove that voluntary actions are initiated unconsciously.[30]

Miller and Trevena explicitly claim that accepting the idea that our decisions are made by our subconscious neuronal processes was pre-

[29] J. Trevena, J. Miller, *Brain Preparation Before a Voluntary Action: Evidence Against Unconscious Movement Initiation*, "Consciousness and Cognition" 2010, vol. 19, p. 454.
[30] *Ibid.*, p. 454.

mature. Their experiments seem to suggest that RP has been mistakenly identified as the real cause of the behaviour while its role serves only to set the brain in attention mode.

Banks and Isham, in their redesigned version of Libet's work, focused on the subject's ability to report the moment of becoming aware of an intention to act. They claim that subjects, rather than perceive the moment of making a decision, infer it based on the content of external input information.[31]

The point of departure was a well-established fact that subjects report a moment of decision making, which happens 300 milliseconds *after* the beginning of brain electrical activity that normally precedes the performance of simple actions. Researchers provided a misleading auditory signal which was played 5 to 60 ms later than the actual movement. The signal was played *after* the performance of an action, just to inform the subjects about their decision. The subjects time of decision moved forward in time linearly with the delay in an auditory beep. If the readiness potential and "awareness of the intent to act" have a rigid connection, then the delayed information about the moment of the action should be irrelevant. However, according to this study, even post-factum information can mediate our ability to introspectively perceive the feeling of making a decision:

> We conclude that a large component, possibly the entire estimate, of W (awareness of the intent to act) is retrospectively inferred from the response, or "postdicted". We do not take our findings to indicate that conscious intention has no role in behavior, but rather that the intuitive model of volition is overly simplistic – it assumes a causal model by which an intention is consciously generated and is the immediate cause of an action.[32]

---

[31] W.P. Banks, E.A. Isham, *We Infer Rather Than Perceive the Moment We Decided to Act*, "Psychological Science" 2009, vol. 20, pp. 17–21.
[32] *Ibid.*, p. 20.

Banks and Isham imply that relying upon the introspective reports of subjects is not only misleading, but incorrect. Their results support the philosophical remarks against Libet's interpretation. Subjects do not report the objective time of a decision making moment, but only what seems to be the time, and Isham and Banks showed that this belief can be easily changed with external information input.

Biological determinism is again at the heart of philosophical discussions. Neuroscientific research devoted to investigating the neuronal basis of human decision making processes provides us with new knowledge that can be interpreted variously. We have tried to show that even in the case of Benjamin Libet's work there are still serious methodological, conceptual and philosophical objections that do not allow us to claim that the deterministic vision of human nature is the correct one. We have to remember that Libet's experiment started this discussion and that it now spans over thirty years and yet the biological determinism thesis cannot be accepted. Neuroscience has put the thesis of biological determinism on the philosophical map again, but it has not made it more reliable than it used to be before the neuroscientific revolution.

2. Behavioural Genetics. Genes – the Ultimate Puppet Masters?

Philosophical discussions over the proper understanding of genetic determinism wasn't so complex (and complicated) as the discussion over free will. Free will is one of the perennial problems in philosophy, while the concept of genetic determinism was developed during second half of the 20th century. It may be better to look for the meaning of genetic determinism in genetics than in philosophy. What we know for sure is that genetic determinism is a form of biological determinism. In the case of genetic determinism, we refer to genes as the ultimate causes of our behavioural and physical traits. We can say fairly that in behavioural genetics we have an unwritten agreement among scientists, who claim that any kind of deterministic interpretations are incorrect. Yet, a very concept of determinism is rarely explored and defined.

Careful analysis of the experiments in genetics allowed Jonathan Kaplan to propose two basic understandings of determinism. The first one he called "complete information" strand, and the second that "intervention is useless".[33] The first understanding of determinism resonates with the folk intuition of the role of genes in the shaping of our nature. It states that everything about us, our physical and behavioural traits, our character, our choices are predictable and programmed by genes. Today we are not so sophisticated with our methods and tools, but in the future we shall definitely see that the paths gene-character or gene-behavioural are simply and causal. Unfortunately for a small group of supporters, this strand of genetic determinism is trivially false from the point of view of contemporary science. We do not need more sophisticated tools, methods, machines, computers and so on to learn what we have already learned. The vast amount of research has proven that the path gene-physical trait or gene-behavioural trait is far from simple and straightforward.

Plomin famously stated that genes are not master puppeteers pulling our strings and we are not just puppets controlled by genetic information.[34] Also Dawkins, who wrote in the *Selfish Gene*[35] that people are robots, machines which are blindly programmed to preserve the selfish replicators known as genes, rejects this radical determinism in genetics. Because this strand of genetic determinism is largely abandoned we will also not investigate it further.

The second "intervention is useless" strand is more subtle and interesting. At the heart of this strand is the idea that any given trait (physical or behavioural) that has genetic etiology cannot be changed or even modified by environmental conditions. If a particular trait is

[33] J. Kaplan, *The Limits and Lies of Human Genetic Research*, Routledge, New York 2000, pp. 9–22.

[34] Cf. R. Plomin et al., *Behavioral Genetics*, W.H. Freeman and Co., New York 1997.

[35] R. Dawkins, *The Selfish Gene*, Oxford University Press, Oxford 2004.

genetic, then a person "has stuck with it".[36] This version of genetic determinism is at the same time far less ambitious than the first one and the more dangerous. It is more dangerous, because, at first glance, it is intuitive and may be difficult to reject. The question of whether this strand is correct is really a question as to what are the limits of genetic explanation? What is genetic information really telling us? How far can we mediate the influence of genes, if at all? These are the most difficult questions to answer, but a few observations have been made which can suggest possible solutions for this puzzle.

The "Intervention is useless" strand means that every trait with a recognized genetic etiology cannot be changed, regardless of the method applied. This understanding of genetic determinism is more interesting than the "complete information" strand also because of its testability. How can we test this strand? If we would like to test it, we need to identify a trait which has genetic etiology. Once recognized, a person with a particular trait can become a subject of "environmental" intervention.

There are several single-gene defects which have been recognized by geneticists. In this paper we would like to present one which serves as a paradigm example of successful "environmental" intervention. The defect in question is phenylketonuria (PKU). A single--gene defect that leads to severe mental retardation. The retardation is caused by a high level of the amino acid phenylalanine that, due to genetic mutation, cannot be metabolized by an organism, which leads to its high level in the blood:

> Children with PKU carry genes that render them unable to metabolize the amino acid phenylalanine. Left untreated, they build up large amounts of pheylalanine in their cells (and lack sufficient amounts of another amino acid, tyrosine) and become severely mentally retarded.[37]

---

[36] Cf. D. Hamer, P. Copeland, *Living with Our Genes: Why They Matter More Than You Think*, Doubleday, New York 1998.

[37] P. Kitcher, *Behind Closed Doors: Junior Comes Out Perfect*, "New York Times" 1996, vol. 29, p. 124.

If the 'intervention is useless' strand is supposed to be the correct one, then children with such an inability to metabolize phenylalanine are doomed to suffer severe mental retardation. Fortunately, this version of genetic determinism is also not true:

> PKU individuals do not suffer retardation if a diet low in phenylalanine is provided during developmental years. Thus, an environmental intervention was successful in bypassing a genetic problem. This important discovery was made possible by recognition of the genetic basis for this particular type of retardation.[38]

Therefore, even when recognized, the single-gene defect can be modified, ameliorated or remediated. In the case of PKU, even a trait with genetic etiology can be changed due to environmental factors, such as, for example, a diet. Jonathan Kaplan referred to the description of the PKU study which can be found in Vigue's *Fear of the Inflexible Gene* in his work:

> It is often assumed that traits with the highest heritabilities cannot be modified by the environment. With our burgeoning understanding of how genes work, this assumption is becoming less and less valid. Consider, for example, the mental retardation associated with PKU (phenylketonuria), a disorder caused by a single defective gene. There is no question that PKU is a classic inherited disorder and that the symptoms will develop in almost any normal human environment. Once the mechanism whereby the PKU gene produces its devastating effects was understood, however, the cure for its symptoms became obvious: PKU is a disorder of phenylalanine metabolism. When phenylalanine is eliminated from the diet of PKU babies, the associated retardation does not develop…[39]

---

[38] R. Plomin, *op. cit.*, p. 9.

[39] L.C. Vigue, *Fear of the Inflexible Gene*, "American Biology Teacher" 1996, vol. 58, pp. 86–88.

Two conclusions could be drawn from this story. Firstly, genes are inherited, but we are not doom to possess the traits associated with them. Secondly, even if we inherit a particular trait, there is a possibility to change it with external (environmental) intervention. Therefore intervention is not useless.

To sum up, we have rejected the first version of genetic determinism, the 'complete information' strand. Next, we have described the PKU story which serves us as a reason for rejecting the second strand of genetic determinism, namely that "intervention is useless". However, many believe that these two forms of determinism are not the only ones that can be identified and accepted in behavioural genetics. People who strongly oppose deterministic claims and do not want to be associated with genetic determinism very often tacitly accept theses that can in fact be deterministic in their nature.

The first example is a widely accepted thesis which states that genetic information is necessary to understand, explain and predict any given trait which has even partially etiology. The second claim is much stronger, since it states that traits even with partial genetic etiologies should be treated as primarily genetic ones. This claim can is often called the "causal thesis". According to it, the expression of genes in the case of a particular trait can be modified *only* through direct intervention.

What follows from these two theses? The first one is just a methodological stance. It says that if one wants to understand the roots of the trait then it *should* focus on genetic roots. If one wants to predict how he will grow old or how his diseases will develop in the future, he *should* turn his attention to genes. And, arguably the most important implication, if one wants to cure his condition he *should* be equipped with genetic knowledge.

We learn from the first thesis that it will be *wiser* to be informed about a trait's genetic etiology, even in cases when we know that this trait, i.e. a disease has only partial genetic etiology. It's very hard to argue against such stance, because when we know that genes affect (or can affect) a trait, then in order to know how this trait came in to existence, it would be good for us to check its genetic etiology. This claim informs us only as to how we should proceed, and does not tell us anything definite about the causal role of genes.

On the other hand, the second thesis can be interpreted as an expression of determinism in genetics. It assumes that between the genotype and phenotype there exists some kind of casual relation. If we encounter normal conditions and are equipped with knowledge about the genotype (and genes expression), then we should be able to predict the structure of any given phenotype. This claim is very often accepted by geneticists, who at the same time reject the "complete information" strand of genetic determinism.[40] This cognitive dissonance is a result of sharing at the same time a very popular belief that there is no more fundamental or basic knowledge than geneticsand, if one is interested in her faith, it should turn to genetics, and conviction that "complete information" strand explicitly stated is trivially false.

Let us re-examine the PKU story. On the one hand, it can be used as an argument for rejecting 'complete information' and 'intervention is useless' strands. On the other, it affirms the casual thesis. It states that traits even with partial genetic etiologies should be treated as primarily genetic. If we want to understand, cure and predict diseases, we should focus on genes. The influence of genes on the development of particular trait can be modified only through direct medical intervention. If the medical interventions are not applied, a human's destiny is determined by genes. An understanding of the path gene-trait is crucial for explaining, predicting and changing traits that have genetic etiology. This claim appears to be justified by such stories as that of PKU.

However, this story has two versions, the first having been told earlir. This version of the story is the most popular in the literature and has been referred to as part of "scientific and clinical folklore".[41]

---

[40] W. Gilbert, *A Vision of the Grail*, [in:] *The Code of Codes: Scientific and Social Issues in the Human Genome Project*, eds. D.J. Kevles, L. Hood, Harvard University Press, Cambridge, Mass. 1992.

[41] Cf. R.M. Murphy, *Phenoylketonuria (PKU) and the Single Gene. An Old Story Retold*, "Behavior Genetics" 1982, vol. 13, pp. 141–157.

The second version of the story is the same story, but much richer in details. One of the aspects of the story not mentioned earlier was the process of the development of the cure. The story begins when two children in Norway were examined by Asbjorn Folling, a doctor, who associated the odour of their urine with a very high level of phenylpyruvic acid. Folling proposed a solution that these high levels were caused by the organism's inability to metabolize phenylalanine. He identified it as an inborn inability connected by an unknown relation with mental retardation. It must have taken over twenty years to develop diets low in phenylalanine that successfully cured this condition and even more time between developing appropriate diets and acquiring knowledge about the genetic basis of this condition. What follows from this story?

Firstly, knowledge about genetic etiology was not available to Folling when he connected the cause of the urine's odour with the inability to metabolize phenylalanine and that these problems might be inborn. Secondly, even if Folling had had knowledge about the genetic roots of PKU condition, it wouldn't have changed the proposed treatment:

> The problems with the treatment, and their solutions, emerged not from subtle excursions into molecular biology and genetics, but from the sorts of pragmatic (mostly trail-and error based) medical research that most often advanced health care.[42]

Therefore, knowledge about the etiology of this illness was not necessary to propose a relevant treatment. In the case of PKU, this knowledge is irrelevant, because with or without treatment it stays the same.

Thirdly, during the time between discovery of the PKU and the development of the right cure, an interesting fact was noted. Not every child with recognized PKU suffered the mental retardation. Some children achieved very good results on their IQ tests even with the

---

[42] J. Kaplan, *The Limits...*, *op. cit.*, p. 18.

PKU condition. This fact stands in opposition to the statement that if one has a genetic defect, then one is stuck with it.

Fourthly, many times in the history of PKU, medical trial-and-error procedure often preceded its scientific phase.[43] It means that scientists have been able to develop a cure for a genetic disease without any knowledge about its genetic basis. Moreover, the way the cure is developed is often of great importance to the process of investigating the roots of the disease. Problems with developing the right treatment were the real source of knowledge about the complexity of PKU while the difficulties with the effectiveness of the treatment in the case of a few patients led to the conclusion that PKU is clinically and genetically heterogeneous.

Clinical heterogeneity is a feature of the genetic condition which describes the situation when the same genetic defect in different people leads to different consequences. Genetic heterogeneity is a situation when different genetic mutations lead to the same illness.

PKU is a genetic condition which can be caused by different genetic mutation.

Let us return to the casual thesis that traits even with partial genetic etiologies should be treated as primarily genetic and the expression of genes for a particular trait can be modified only through direct intervention. What does the second version of the PKU story tell us? Is the casual relation between genotype-phenotype really so obvious? Not only does the second PKU story deny "complete information" and "intervention is useless" strands, but also it possesses serious difficulties for the casual thesis. Folling did not need the subtle knowledge from molecular biology and genetics in order to connect children's illness with problems with their metabolism. Moreover, the belief that these problems are somehow inborn has not resulted from genetic data. Someone may say that he was just lucky and his lucky guess cannot be an argument against casual thesis. It would be true, if the other facts about PKU had not been discovered. The development

[43] *Ibid.*, p. 18–19.

of the appropriate cure took over 20 years to be completed. During this time, scientists acknowledged that the process of making a cure for PKU had not been aided by genetics. It means that the conclusion which is often drawn from the causal thesis that genetic knowledge is necessary to explain, predict and mediate illness with genetic etiology is simply not true (or at least not always true). In some cases it is even irrelevant.

Another conclusion which follows from the casual thesis states that the process of gene expression in normal circumstances is a kind of casual process which means that we simply have a casual relation between genotype-phenotype. This trivial and simplistic approach to genetics is also radically untrue. Clinical and genetic heterogeneity in the clearest way possible tells us that even with a genetic condition that causes PKU illness and then mental retardation is not always necessary for mental retardation to appear. Children with a diagnosed PKU illness, even without direct medical intervention, do not suffer from it and different patients show different symptoms. Moreover, different genetic mutations can result with the same phenotypic traits, such as PKU illness. Therefore the relation between genotype-phenotype could be far from a causal one.

As we have seen, genetic determinism is a form of biological determinism. We have presented the PKU story in order to show that there is no real reason for accepting genetic determinism in any form. What is more important, we do not need to introduce philosophical arguments against a deterministic view of human nature. We simply have to refer to genetic research. This method has shown that in order to reject genetic determinism we cannot ignore any fact with regards to such studies as the PKU story. Proponents of genetic determinism chose to ignore scientific facts and then interpret what is left in order to justify their own philosophical position. The main enemy of genetic determinism seems to be genetics itself.

3. From Neuroscience and Genetics to Neurogenetics

Nowadays, biological determinism claims can also be assessed from the neurogenetic point of view. Neurogenetic studies focus on the role of genetics in the development and function of the nervous system.[44] It combines our knowledge form neuroscience and genetics. We can track the development of any given organism from its birth through life to its death. This kind of study allows us to understand the role of genes in the development of any given organism and the role of the nervous system in shaping the behavioural traits of such organism. Unfortunately, it is far too early to conduct such ambitious studies on humans. We still have much to learn when it comes to understanding genetic influence on our development and how genes can affect our nervous system, and especially, our brains.

However, there is no (urgent) need to understand such complex processes by studying humans. In biology one can find very interesting type of entities, called model organisms.[45] The main thought that lies at the bottom of the idea of such organisms is that certain entities can be carefully and extensively studied and the results of such a study can be extrapolated from the model entity to more complex ones. A typical model organism has a short life cycle, a small genome with respectively large chromosomes, a small size and its development must be complex enough to provide information about a similar biological process in more complex organism. We can say that the model organism serves as a reference point for far more developed ones. In other words, when we understand how genes and the nervous system affect behaviour in such simple organisms, we can use it as point of departure to understand how these same processes work in the case of humans. If a model organism is fixed by its genes and its nervous system, it may be the case that humans are fixed and programmed.

---

[44] P. Greenstein, T.D. Bird, *Neurogenetics. Triumphs and Challenges*, "West Journal of Medicine" 1994, vol. 161, pp. 242–245.

[45] Cf. S. Fields, M. Johnston, *Cell Biology. Whither Model Organism Research?*, "Science" 2005, vol. 307, pp. 1885–6.

C. elegans, or as Robert Cook-Deggan[46] referred to 'a reductionist's delight', is the model organism which is important from the point of view of the paper. It has been studied since 1870s, but it has been at the centre of research in developmental biology since 1960s. Today we have almost everything about its development, the role of its genes and nervous system. According to Schaffner,[47] the interplay between genes, the nervous system and environment is best understood in the case of *C. elegans*. Today we know even how each cell arrived at its destinations, including which cells die during *C. elegans'* development. We know not only the genome of *C. elegans*, but also its connectome,[48] a comprehensive map of the neural connections in the nervous system, a complete reconstruction of all its neural and synaptic connections.

It is far easier to investigate how 'all-mighty' biological factors affect behaviour in a nervous system while studying simpler organisms. If we want to investigate whether deterministic claims are backed by scientific data, a good idea would be to seek answers there where one can find as much relevant knowledge as possible. Therefore today, we should pay closer attention to one model organism, namely, *C. elegans*.

In its case, the relationship between genes, the nervous system and the environment is ruled by seven default rules. The first four focus on the interplay between genes and neurons, the further two refer to environmental factors and the last one takes us back to genetics.

1) 'many genes – one neuron' – coming into existence and development of one neuron can be affected by many different

---

[46] Cf. R. Cook-Degan, *Gene Wars*, Norton, New York 1994.

[47] K.F. Schaffner, *Complexity and Research Strategies in Behavioral Genetics*, [in:] R.A. Carson, M.A. Rothstein, *Behavioral Genetics. The Clash of Culture and Biology*, The Johns Hopkins University Press, London 1999, pp. 61–88.

[48] Cf. J.G. White, E. Southgate, J.N. Thomson, S. Brenner, *The Structure of the Nervous System of the Nematode Caenorhabditis elegans*, "Philosophical Transactions of the Royal Society B: Biological Sciences" 1986, vol. 314, pp. 1–340.

genes, this rules is sometimes called "statistical epistais" rule';

2) 'one neuron, x neuronal systems, x behaviours – many neurons can affect many different behavioural traits;
3) 'x neuronal systems, one behaviour' – many neuronal systems can influence one behavioural trait;
4) 'one gene, many behavioural traits' – pleiotropic genes affect many phenotypic traits, including behaviours;
5) 'development is stochastic' – during development different nervous systems can be created;
6) 'phenotypic plasticity' rule – diverse developmental environments cause diverse behavioural traits in genetically identical organisms;
7) 'physical epistatis' rule – genes are often interact with each other in order to shape the development of particular neurons or systems of neurons.[49]

What is the philosophical meaning of this set of rules? Firstly, we will explain the biological consequences of them, and then we will propose a philosophical interpretation of the biological data. Our point of departure is a well-known and widely accepted rule that states that one neuron can be influenced by multiple genes. The first rule states that a particular set of genes can be responsible for the coming into being and development of one neuron. The fourth rule refers to the specific feature of genes which can affect the nervous system. This feature is called pleiotropiy. The mutation of any given pleiotropic gene can result in changes of many phenotypic traits, such as behavioural traits. It means that there is a possibility that many behavioural variations are caused by genes which are normally not recognized as being responsible for such behaviour. We have to remember that our behaviour, as we have seen earlier, is a final product of our brains, which consists of a great number of neurons.

[49] K.F. Schaffner, *Complexity...*, *op. cit.*, pp. 74–75.

The second and third rule introduce two other interesting features. The second one states that neurons don't affect our behavioural traits separately, more often than not, they are formed in a system – a network of neurons processing one information type. The working environment for one neuron are other neurons which are organized in the network. Moreover, a particular system, even when recognized as responsible for a certain behaviour, can also be involved in other behaviours. This means that one neuronal system can affect many various behaviours.

The relation between genes, the nervous systems and behaviour does not tell us the whole story. We said that we have different types of genes, but there are also different types of neurons. One which is of great importance for us are multifunctional neurons. The mulitfunctionality features of neurons pose a serious difficulty for anyone who wants to easily jump to conclusions while interpreting the activity of human brains. Multifunctionality means that any given neuron can be a part of a large number of neuronal systems, therefore its activity can affect many types of behaviour.[50] For example, when we can identify a particular neuron which is active during the performance of an experiment, it does not mean that this neuron (or neuronal system) is definitively responsible for performing a particular action. Maybe it is involved in other activity which appeared, but was ignored during the experiment. Moreover, when we successfully discover that a particular neuronal system influences a certain type of behaviour it doesn't mean that this relation "neuronal system – behaviour" is an exclusive one.

Let us stop here and try to sum up. The first four rules, as we said, describe the relation "genes-neurons-behaviours". This relationship is far from simple and straightforward. Its complexity grows, especially, when one takes under consideration the mulitfunctionality of some neurons and the pleiotropic genes. Due to this difficulty, it is al-

---

[50] P.S. Churchland, T. Sejnowski, *The Computational Brain*, MIT Press, Cambridge 1992, p. 349.

ready very hard to track the influence one gene can have on particular behaviour, even in the case of C. elegans. In the case of humans, it is almost impossible. Nevertheless, other factors also have to be taken under consideration.

Fifth rule, "development is stochastic", in contrast to the preceding four, refers to a non-biological factor. The reason for introducing this rule was a suprising fact that two genetically identical specimens of *C. elegans* bred in an identical environment have different synaptic connections. In other words, their nervous systems differ. Scientists have proposed three alternative scenarios that provide three varying answers to this puzzle.[51]

According to the first one, the cause of such a difference is purely genetic in its nature. The differences in synaptic connection are a result of hidden, unrecognized genetic mutation. Therefore, if we were able to recognize these mutations today then we would be able to track the influence of them on differences in synaptic connections. But this scenario is very hypothetical. We do not know for certain if such mutations have actually happened, and we are just not technologically prepared to recognize them. The "genetic" scenario is an interesting one, because it can been seen as an expression of a belief that everything is programmed and controlled by biological factors, in this case, by our genetic structure.

The second scenario takes an opposite stand and it looks for the causes of such differences between two specimens of C. elegans not in their genes but in the environment. The differences are the results of adaptation processes to different environmental conditions. Even in a laboratory there is no possibility of creating two, identical environments. Even if the differences are minimal, there are, nevertheless, variations. As a result we have different C. elegans, developed in a different environment, through mechanisms of slightly different adaptation processes. This "environmental" scenario is very interesting, because it puts an accent on the role of the environment in shaping

[51] K.F. Schaffner, *Complexity*..., *op. cit.*, pp. 74–75.

our nervous systems and, ultimately, our behavioural traits. The consequences of accepting this stand would be as follows: genes are not fully in control of us, or even any other (and much simpler) organisms for that matter. Every theory which refers only to a genetic explanation when it comes to our behaviour must be, in principle, an insufficient one.

The "environmental" explanation of differences in the structure of the nervous systems of two identical genetic specimens of C. elegans, if true, has very surprising consequences. Even the slight difference in the (laboratory) environment during development matters. Therefore, in real life, where differences between developmental environments are almost immeasurable, the recognition between genetic and environmental causes can be very difficult to conduct. However, this scenario is still untested and it serves only as a "fulfilling-gap-hypotheses". Scientists are not sure if the different condition actually happened in the laboratory and, therefore, the "environmental" scenario is as hypothetical as the "genetic" one.

The last scenario is – without doubt – the most honest and realistic one. It states that the developmental process which is controlled by our genes and under the influence of the environment we grow up in, is stochastic in its nature. It means that two identical genetically specimens which arise in the same environment will have different synaptic connections and therefore different nervous systems which can result in different behavioural traits due to the existence and influence of stochastic noise. The stochastic process, due to stochastic noise, is a process whose subsequent state is a result, on one hand, of the rules that have governed the process and, on the other hand, of random elements in developmental process which are called stochastic noise.[52]

To sum up, full knowledge about genetic structure and the environmental conditions is not enough to predict the developmental process from its beginning to its end. The stochastic noise changes the

[52] *Ibid.*

ways of interacting genes with environments and, perhaps, the process of gene expressions.

The sixth 'phenotypic plasticity' rule states that diverse developmental environments cause diverse behavioural traits in genetically identical organisms. This rule makes us of the plasticity phenomenon. Plasticity (or synaptic plasticity) is a feature of a synaptic connection that allows them to change throughout their life. The strength in response of a synaptic connection between two neurons can be modified due to their use or disuse. It means that, even when we have a fully developed nervous system, there is still room for modifying it under the influence of the external environment. These changes can be seen outwardly as changes in behavioural traits. Studies conducted on C. elegans' behaviour have led to the conclusion that even in that simple organism, the plasticity of the synaptic connection is present and can have an important influence on the behaviour of the worm. Two kinds of plasticity have been identified while observing the behaviour of C. elegans: short-time adaptation and learning.[53]

The worm has been exposed to different environmental conditions which have caused a completely different set of behaviour. For example, when deprived of food and put into a cold environment, a worm which normally lives alone, has now started to form groups and behave more energetically. When the external condition were returned to the initial state, the behaviour of the worm was again normal. Thanks to this experiment, a very convincing explanation was proposed. Different environments force the worm to develop novel behavioural traits which are correlated with changes in their synaptic connections. These changes can be permanent (in that case we are dealing with learning) or short-time (in this case we are concerned with short-time adaptation) and they are introduced with little regard to genetic predispositions.

The last, seventh rule takes us back to genetics. However, in opposition to the first four which describe the relationship between

[53] *Ibid.*, p. 74.

genes and neurons, the last one focuses on the genetic environment. What does this mean? Genes function in a genetic neighbourhood and any given gene, in order to affect our development, must interact with other genes. However, we do not have only one type of genes (i.e. pleiotropic genes). Some genes are "equipped" with tools that allow them to modify the activity of other "unarmed" genes. Epistasis means that the effects of one particular gene are modified by one or several other genes. We also have suppressor genes that suppress (block) the phenotypic expression (i.e. a particular behavioural trait) of another gene.

Physical epistasis is really disturbing for anyone looking for a simple answer about biological factors in genetics to matters describing our behaviour. For example, assume that we have found a particular genetic mutation which can be seen as the most primitive cause of our antisocial behaviour. Does it mean that every time when spotted in the phenotype of any given human it will "make" him behave in that way? No, for many reasons, but among others because other mutations can happen (which we are not aware of for now) and modify the expression of the antisocial gene or even, as in the case of suppressor gene, the expression of it will be cancelled.

Neurogenetic is a new kind of biological discipline, but its achievements could be of great importance to anyone who is interested in investigating the role of biological factors in an organism's behavioural traits. Today we can learn from neurogenetic studies that these studies do not support any form of biological determinism. Studies on such a simple organism as *C. elegans* have shown that the relation between genes, nervous systems and environments are far from the simplicity of biological determinism. Even in the case of a 1 mm. nematode, we are not able to point to a relation between biological factors and behavioural traits that can be roughly described as casual.

Conclusion

Biological studies do not lead to accepting the thesis of biological determinism. We have shown that even in paradigm cases in neurosciences, behavioural genetic and neurogenetic proponents of biological determinism over-simplify the results of scientific research. We believe that the main weakness of such studies is the poor interpretation of them which leads to insufficiently grounded philosophical conclusions. Moreover, as we have seen in the case of Libet's work, authors do not offer an explanation for the methodology applied and also overlook the philosophical assumptions made by themselves when designing the experiments.

Neurogenetics offers the most complete and complex picture of the impact of biological factors on the physical and behavioural traits of organisms. Neuroscience and behavioural genetics are both more entangled in philosophy than neurogenetics.[54] We believe that studies on model organisms can tell us far more about the extent of the impact of biological factors than we can learn today from neuroscience and behavioural genetics.

We believe that neurogenetic studies on model organisms should be a point of departure for anyone interested in identifying the role of genes, nervous systems and environments in determining our behaviour. Studies on one millimetre nematode *C. elegans* reveals far more information than Libet's studies devoted to simple motor movement or studies on one gene-defect called PKU. Moreover, nothing in neurogenetic studies allow the formulation of such a simplistic and naive thesis that biological factors determine our choices, behaviour and even our lives. The story of biological determinism should be retold.

[54] When it comes to philosophical assumptions in neurosciences cf. B. Brożek, *Normatywność prawa*, Wolters Kluwer business, Warszawa 2012, pp. 182–195.

Wojciech P. Grygiel
The Pontifical University of John Paul II
Copernicus Center for Interdisciplinary Studies

Quantum Mechanics and Its Role in Cognitive Sciences
A Critical Survey[1]

Introduction

The quest for a theory that would comprehensively describe the workings of the human mind has attracted the great attention of researchers since time immemorial. Although the biology of the brain is now very well understood, the main question hinges upon the explanation of the mental phenomena traditionally reserved to the activity of the immaterial mind. This is inconsistent with any type of the mind – body dualism in which the immaterial mind exerts its causal influence on the biological tissue of the brain. In other words, no laws of science can relate the material to what is immaterial. The dualist stances are no longer considered as viable proposals in favour of the functionalist approaches where the workings of the mind are regarded as the function of the brain. This makes the formulation of the theory of the mind a plausible task. Daniel Denett encapsulates this well by stating:

> a cobbled-together collection of specialist brain circuits, which . . . conspire together to produce a . . . more or less well-designed

[1] This paper was written within the research grant "The Limits of Scientific Explanation" sponsored by the John Templeton Foundation.

> virtual machine . . . By yoking these independently evolved specialist organs together in common cause, and thereby giving their union vastly enhanced powers, this virtual machine, this software of the brain, performs a sort of internal political miracle: It creates a *virtual captain* of the crew . . .[2]

The main difficulty, however, is how one should translate the mental qualities (qualia) such as sound, colour, pain, pleasure and others into the scientifically manageable language. Moreover, such classically defined notions as understanding, consciousness, free will become a greater challenge. It is commonly believed that these qualities can ultimately be explained by identifying and formulating the dynamics of the corresponding neuronal structures.[3]

The chief aim of this paper is the review and the critical assessment of the application of quantum mechanics in the explanation of the mental phenomena. Consequently, much less attention will be paid to the accuracy of the notions characterizing the mental phenomena specified above, especially that their precise meaning may vary in different cases especially that the proponents of the discussed stances are of various level of expertise in cognitive sciences themselves. Be that as it may, quantum mechanics turns out to be attractive to researchers in the area of the cognitive sciences for two reasons. These reasons are: (1) the global character of the quantum mechanical wave function potentially suitable to for the explanation of conscious phenomena extending over large areas of the brain, (2) the quantum mechanical indeterminism based on quantum probability that is expected, for instance, to facilitate the unconstrained transposition of the notion of free will into the scientific terms. Literature offers several broader reviews on the motivations and ways

---

[2] D. Dennett, *Consciousness Explained*, Little, Brown and Co., Boston, New York, London 1991, p. 228.

[3] Cf. P.S. Churchland, *Brain-wise: Studies in Neurophilosophy*, MIT Press, Cambridge, Mass. 2002.

in which quantum mechanics is used in the explanation of the mental phenomena.[4]

Before any analysis is undertaken, however, it must be remembered that as soon as quantum mechanics was formulated in the 1920's, it provoked a major discussion on the role of the observer in the quantum experiment.[5] This is mainly the result of the dependence of the measurement's outcome on the act of the measurement itself. Possible interpretations skyrocketed into claims that it is actually the human consciousness that causes the famous collapse of the wave vector thereby, for instance, deciding on the fate of the Schrödinger Cat.[6] Interestingly enough, quantum mechanics turned out to support the dualist stances in regards to the mind-body problem mainly as a consequence of the radical incompatibility between the quantum – mechanical formalism and the data perception exclusively in classical terms. Since there is no common agreement on the interpretation of quantum mechanics and its precise physical meaning is controversial, the application of the theory to the explanation of mental phenomena must be carried out with great caution.[7]

The corpus of the paper opens up with a brief exposition of the mind – relevant quantum mechanics with the aim to explain the aspects of the theory's formalism that are most frequently invoked as the mental phenomena acquire their quantum-based explanations.

---

[4] Cf. Ch. Koch, K. Hepp, *Quantum Mechanics in the Brain*, "Nature" 2006, vol. 440, pp. 611–612; M. Lockwood, *Mind, Brain and the Quantum,* Basil Blackwell Ltd., Cambridge, Mass. 1989.

[5] An excellent study of Niels Bohr struggles with the explanation of the relation between a quantum measurement and an observer can be found in: D. Murdoch, *Niels Bohr's Philosophy of Physics*, Cambridge University Press, Cambridge 1987.

[6] F. London, E. Bauer, *La Teorie de l'Observation en Mécanique Quantique*, Hermann & Cie., Paris 1939. A good exposition of London and Bauer's ideas can be found in: M. Jammer, *The Philosophy of Quantum Mechanics*, *The Interpretations of QM in Historical Perspective*, John Wiley and Sons 1974, pp. 482–486.

[7] For a review of the contemporary interpretations of quantum mechanics with a philosophical commentary see: W.P. Grygiel, *Interpreting Quantum Mechanics: Why An Interpretation?* [in:] *Wyzwania racjonalności: Księdzu Michałowi Hellerowi współpracownicy i uczniowie*, eds. S. Wszołek, R. Janusz, Wydawnictwo WAM, Kraków 2006, pp. 113–131.

Some brief remarks are made concerning the sizes of the atomic and molecular objects that are thought of as being relevant at the level of the neuronal activity. Following this, the four best known proposals of how quantum mechanics can be utilized in such explanations are discussed in the following order: Roger Penrose (with Stuart Hameroff), John C. Eccles, Henry Stapp and David Chalmers. The reason for such a sequence of presentation is the degree to which the authors "stretch" the physically meaningful formalism of quantum mechanics to fit their philosophical desiderata. David Chalmers, for instance, goes as far as to attempt to relieve the problems of quantum mechanics with his preconceived theory of mind. Although these proposals seem quite divergent at face value and they are put forward by researchers of varying provenience, each of them ultimately supports a specific form of dualism. It is also quite evident that if the philosophical imagination is pushed too far, an unscientific form of pseudo-mysticism may easily arise.

1. Mind – Relevant Quantum Mechanics

Quantum mechanics is an elaborate physical theory that was formulated with the aim of providing a description of the atomic phenomena.[8] The basic means by which the theory gives an account of physical states is the so called wave vector ψ that deterministically evolves in time according to the time-dependent Schrödinger equation:

$$ih\frac{d\psi}{dt} = \hat{H}\psi$$

In other words, unperturbed quantum systems behave identically to classical systems in terms of their unique time evolution. The pro-

[8] Any college textbook can be consulted for theoretical foundations of quantum mechanics. For an introduction with a good structural basis and elements of philosophical commentary see: C. Isham, *Lectures on Quantum Theory: Mathematical and Structural Foundations*, Imperial College Press, London 1995.

babilistic character of quantum mechanics is revealed *only* in the process of measurement performed on a quantum system. If the system is described by a superposition of wave vectors as prescribed by the following equation:

$$\Psi = \sum_i c_i \psi_i$$

the experimental procedure leads to the collapse (or reduction denoted as **R**) of the wave vector Ψ . As a result, a single number is obtained with the probability proportional to $|c_i|^2$ that corresponds to the eigenvalue λ_i of a quantum mechanical operator $\hat{A}$ satisfying the equation:

$$\hat{A}\psi_i = \lambda_i \psi_i$$

As far as the two main features of quantum mechanics that are of potential interest for a cognitive scientist, the theory is precise and straightforward. The wave function is indeed of a global character because its dependence on the spatial parameter *x* is known for all values of *x* of the set of the real numbers indicating that a state of an unperturbed quantum object such as a particle is not bound to a spacetime point corresponding to the particle's location detected in a measurement. There exists a vast literature on the problem of testing the so called Bell's inequalities implying the non-local character of quantum phenomena proper to non-interacting particles such as photons.[9] Moreover, contemporary efforts to formulate the theory of quantum gravity suggest that non-locality maybe a fundamental property of the physical reality at the Planck level.[10] Inasmuch as the non-locality seems to be now more than a mere scientific hypothesis, quantum

---

[9] For a good survey of Bell's Theorem and its philosophical implications see: J.T. Cushing, E. McMullin, eds., *Philosophical Consequences of Quantum Theory: Reflections on Bell's Theorem*, University of Notre Dame Press, Notre Dame, IN 1989.

[10] Cf. M. Heller, L. Pysiak, W. Sasin, *Conceptual Unification of Gravity and Quanta*, "International Journal of Theoretical Physics" 2007, vol. 46, pp. 2494–2512.

states undergo a very rapid process of decoherence as they are exposed to the interaction with the macroscopic environment.[11] In a nutshell, decoherence leads to the quenching of quantum interferences and the system behaves as a statistical ensemble of disentangled constituents.

Since the presence of quantum effects depends on the scale of the objects under study, it now seems fitting to compare the orders of magnitude that are involved in neurobiology (all distances are rounded up to their orders of magnitude only). To make the scale complete, the non-local effects at the Planck level refer to distances of 10^{-34} m, the classically computed size of the electron is on the order of 10^{-15} m, the radius of the atom of carbon is 10^{-10} m, the average protein molecule is 10^{-8} m and the average size of the neuron is 10^{-5} m. The actual founders of the theory of decoherence, Erich Joos and Dietrich Zeh, proceeded to construct a simple model to allow the calculation of distance and time scales at which the process of decoherence is operative.[12] Strikingly enough, they discovered that the quantum interference between two different regions a piece of dust of 10^{-5} cm in radius suspended in the air will be quenched for a distance of 10^{-13} cm on the time scale of nanoseconds. These numbers are additionally corroborated by the calculations performed by Max Tegmark who confirmed that the processes of decoherence in the brain would be too efficient to permit the stability of quantum states occupying larger portions of the brain's tissue.[13]

With these data at hand, it seems rational to suggest that the use of quantum mechanics for the explanation of any mental phenomena will face a great difficulty mainly due to the fact that even on the scale of single neurons, the quantum effects may be extremely short lasting

---

[11] For a general introduction into the theory of decoherence see: R. Omnés, *The Interpretation of Quantum Mechanics*, Princeton University Press, Princeton, NJ 1994, pp. 196–234.

[12] E. Joos, H.D. Zeh, *The Emergence of Classical Properties through Interaction with the Environment*, "Zeitschrift für Physik B" 1985, vol. 59, pp. 223–243.

[13] M. Tegmark, *Importance of Quantum Decoherence in Brain Processes*, "Physical Review E" 2000, vol. 61 (4), pp. 4194–4206.

to permit the buildup of any delocalized states of macroscopic relevance. It must also be remembered that neurons are anything but isolated pieces of dust and their complex interactions with the entirety of the neuronal structure may additionally decrease the estimated characteristic times of quenching of the quantum interferences. Based on these considerations it is now worth taking a closer look at several quantum models of mind to better understand their foundations and credibility.

2. Penrose and Hameroff

It remains beyond doubt that Roger Penrose's expertise in theoretical physics is incommensurate with any of the three other figures presented in this study. Moreover, Penrose has been quite vocal with his philosophical motivations in the theoretical research where the ontology of the three worlds – mathematical, physical and mental plays the central role. A vast amount of literature has been devoted to the study and critical appraisal of his largely speculative proposal.[14] Its philosophical motivations are best spelled out in the opening chapters of the monumental work entitled *The Road to Reality*.[15] Since even a fragmentary repetition of Penrose's ideas would be superfluous, this short survey is intended to put them in somewhat different perspective coherent with the scope of the article.

Firstly, it must be pointed out that Penrose's research is largely focused on formulating the theory of quantum gravity, allowing for the unification of quantum mechanics and the general theory of relativity. In other words, Penrose ventures out to search out for clues facilitating the unification and, contrary to mainstream research, he claims that it is quantum mechanics and not the general theory of

[14] The main response in a form of a prolonged discussion of Penrose's ideas was published in the internet journal "Psyche", 1995–1996, vol. 2 (<http://www.theassc.org/journal_psyche>; accessed on June 6, 2012).

[15] R. Penrose, *The Road to Reality*, Alfred A. Knopf, New York 2005.

relativity that needs revision to permit their final unification.[16] Moreover, this unification will also involve the theory of the mind. It is rarely stressed, however, that the often repeated likeness of Penrose's thinking to that of Albert Einstein consists in Penrose's particular attention to some obvious facts of nature as is was the case with Einstein's elevator.[17]

In regards to the theory of mind, it is his observation that the human mind operates in a non-algorithmic mode. In particular, by applying Gödel's famous Theorem to the non-algorithmicity of mathematical thinking, Penrose extends it to claim that the conscious human mind operates in the non-algorithmic mode in its entirety. Since in his stance on the mind-body problem Penrose is a physicalist, the non--algorithmic character of the human mind necessitates that physics responsible for mind's activity is also of non-algorithmic character.[18] This is achieved by proposing the objective character of the reduction of the wave vector **OR** that according to the yet non-existing physics should also be non-algorithmic. In distinction to the routine reduction of the wave vector arising in the process of a measurement, however, the objective process may occur on its own due to differences in spacetime curvature within the quantum superposition.

In order to develop the neurobiological foundation for his conjectures, Roger Penrose teamed up with a neuroscientist and anaesthesiologist, Stuart Hameroff. Following the publication of Penrose's famous *The Emperor's New Mind*[19], Hameroff suggested that microtubules that provide support for the neuronal structure could become a basis for the existence of entangled quantum states extending over significant portions of the brain and control the functioning of the synaptic connections. Leaving more biological details to the side, the up-

[16] *Ibid.*, pp. 7–23.

[17] R. Penrose, *Singularities and time-asymmetry*, [in:] *General relativity: An Einstein Centenary Survey*, eds. S.W. Hawking, W. Israel, Cambridge University Press, Cambridge 1979, p. 582.

[18] R. Penrose, *The Large, The Small and The Human Mind*, Cambridge University Press, Cambridge 1997, pp. 93–139.

[19] *Idem*, *The Emperor's New Mind*, Oxford University Press, Oxford 1985.

shot of the joint work is that consciousness arises as the result of the orchestrated objective reduction (orch-**OR**) of such quantum states within the brain.[20] As it has been already mentioned, the entire Penrose-Hameroff proposal has been subject to an avalanche of criticism, especially in regards to the stability of quantum superpositions occurring in the brain tissue as they would quickly decohere and have no relevance for the functioning of the brain (e.g., Tegmark mentioned in the earlier portion of the text).

Two points are fitting at this stage. Firstly, the application of quantum mechanics by Penrose and Hameroff mainly relies on Penrose's intuitions in regards to his program of formulation the theory of quantum gravity. Considered highly controversial due to Penrose's insistence on the incompletedness of quantum mechanics, the unification framework involving the twistor theory has not achieved the desired result. Although carefully physically motivated by Penrose, the suggested "stretch" in quantum mechanics itself is still a missing piece of physics, resulting in the proposed theory of consciousness remain in the sphere of sheer hypothesis and speculation. Secondly, Penrose and Hameroff do not escape from the necessity to justify how the conscious experience is linked to neuronal activity. This is accomplished based on their argument that correctly functioning cytoskeleton is responsible for conscious behaviour as evidenced by Hameroff's anaesthetic studies.[21]

This issue receives the most thorough treatment in the recent work of Penrose and Hameroff where, besides a detailed recapitulation of their proposal, they attempt a somewhat risky explanation why consciousness plays central role in the universe.[22] The main premise

---

[20] *Idem*, *Shadows of the Mind: A Search for the Missing Science of Consciousness*, Oxford University Press, Oxford 1994, pp. 349–391; S.R. Hameroff, *Quantum Computation In Brain Microtubules? The Penrose-Hameroff "Orch OR" model of consciousness*, "Philosophical Transactions Royal Society London (A)" 1998, vol. 356, pp. 1869–1896.

[21] S.R. Hameroff, *The Entwined Mysteries of Anesthesia and Consciousness*, "Anesthesiology" 2006, vol. 105(2), pp. 400–412.

[22] R. Penrose, S.R. Hameroff, *Consciousness in the Universe: Neuroscience, Quantum Space-Time Geometry and Orch OR Theory*, "Journal of Cosmology" 2011, vol. 14.

draws upon the fact that the objective reduction of the wave vector **OR** may hypothetically occur anywhere else in the Universe thus making the conscious phenomena much more fundamental to the nature of the Universe itself. The most appealing consequence is that the conscious life could emerge in other parts of the Universe but since Penrose and Hameroff's idea of consciousness and life in general is utterly reductive, this proposal must remain in the zone of sheer science fiction. After all, such may be the authors' intention as they write in conclusion of the paper:

> Nevertheless, sentient creatures might have evolved in parts of the universe that would be highly alien to us. One possibility might be on neutron star surfaces, an idea that was developed ingeniously and in great detail by Robert Forward in two science-fiction stories (Dragon's Egg in 1980, Starquake in 1989). Such creatures (referred to as 'cheelas' in the books, with metabolic processes and OR-like events occurring at rates of around a million times that of a human being) could arguably have intense experiences, but whether or not this would be possible in detail is, at the moment, a very speculative matter. Nevertheless, the Orch OR proposal offers a possible route to rational argument, as to whether life of a totally alien kind such as this might be possible, or even probable, somewhere in the universe.[23]

3. Eccles and the Psychons

As it has been mentioned in the introduction, the enthusiasts of quantum mechanics having an important role in the explanation of the mental phenomena are opponents of the physicalist (materialist) replies to the mind – body problem. The author of the second "quantum proposal" presented in this paper, John Eccles, gave a clear expres-

[23] *Ibid.*

sion to such a stance named *dualist-interactionism* in his work written together with Karl Popper entitled *The Self and Its Brain.*[24] In another influential work of Eccles one reads:

> I maintain that the human mystery is incredibly demeaned by scientific reductionism, with its claim in promissory materialism to account eventually for all of the spiritual world in terms of patterns of neuronal activity. This belief must be classed as a superstition... we have to recognize that we are spiritual beings with souls existing in a spiritual world as well as material beings with bodies and brains existing in a material world.[25]

Each dualist solution of the mind – body problem must ultimately face the difficulty of linking the psychical (mental) regime with the physical. In other words, a unit of *conscious experience* defined usually in common sense terms is correlated with a biological activity of the brain in order to establish a causal connection. Eccles states clearly:

> The new light on the mind-brain problem comes from the hypothesis that the non-material mental events, the World 2 of Popper, relate to the neural events of the brain (the World 1 of matter and energy) by actions in conformity with the physics of quantum theory.[26]

Clearly then, quantum theory is used as the mind-body link. Eccles makes the fundamental assumption that the mental regime is constituted by a kind of a probabilistic field that acts upon the tissue of the brain:

---

[24] K. Popper. J.C. Eccles, *The Self and Its Brain: An Argument for Interactionism*, Springer, Berlin 1977.

[25] J.C. Eccles, *Evolution of the Brain: Creation of the Self*, Routledge, London 1989, p. 241.

[26] *Idem*, *How The Self Controls Its Brain*, Springer, Berlin–New York 1994, p. 56.

> The hypothesis of mind-brain interaction is that the mental events act by a quantal probability field to alter the probability of emission of vehicles from the presynaptic vesicular grids.[27]

How is it accomplished on a biological level? Eccles singles out the so called *dendrons* that are the neural units of the cerebral cortex. Dendrons are in turn connected with the *psychons* that represent the units of the conscious experience. As the brain exercises a cognitive activity such as willing or thinking, a psychon acts on a corresponding dendron leading to the increased probability for a neuron to fire. This, in turn, is effected by quantum tunneling in the process of synaptic exocytosis that is a quantum phenomenon occurring in the presynaptic vesicular grid aiming at the exchange of synaptic transmitters. In order to assure that no violation of the energy conservation laws, Eccles stipulates that the interaction between brain and mind "can be conceived as a flow of information, not of energy".[28] This is controversial, for each exchange of information is ultimately bound with the exchange of some energy-mass equivalent. Since Eccles clearly correlates conscious phenomena with mental intentions, his ultimate link between the mental and the physical is achieved as the voluntary actions arise due to momentary jumps of the probability of exocytosis in a large number of synapses. Aside from matching the mental regime with the entirely enigmatic quantal probability field, Eccles does not stipulate any entangled quantum states that could reflect a neuronal activity extending over a larger area of the brain. His dualistic stance is justified by the clear conviction that the natural sciences are reductive and thus incapable of explaining the complexity of human nature.

---

[27] *Ibid.*, p. 69.

[28] *Ibid.*, p. 9.

4. Henry Stapp and the Old Fashioned Quantum

The application of quantum mechanics to the explanation of mental phenomena put forward by Henry Stapp is a rather elaborate amalgamate of well assimilated physics with an eclectic panorama of philosophical inspirations. This is evidenced in two of his most famous books entitled *Mind, Matter and Quantum Mechanics*[29] and *Mindful Universe: Quantum Mechanics and the Participating Observer*[30]. In both of these he presents a highly original proposal reaching deep into the philosophical foundations of quantum mechanics of how one should relate consciousness to the activity of the brain. His first assumption is that the classical physics is irrelevant to the brain processes and one must use quantum mechanics in order to explain these processes. In his considerations Stapp relies heavily on the early developments of quantum mechanics where it was often maintained that the conscious observer plays an active role in the process of measurement. Stapp often appeals to the works of John von Neumann, with particular emphasis on the postulated process of the wave function reduction. His main goal is to provide a theoretical framework of establishing causal relationships between the mind and the brain attesting to his strong dualist inclinations. In order to achieve that, however, he deeply redefines the underlying ontology so that the desired link of the psychological with the physical can be achieved. Since Stapp's ideas are quite complex, only the main tenets will be signalled.

First of all, Stapp considerably redefines on what is the focus of science in general. Stapp claims that the physical universe is not causally closed and, by shifting the focus of science on how the experimental data are actually sensed, this gap maybe sealed. Thus Stapp establishes the causal link between the mental (psychological) and the physical. For him, quantum mechanics is not the theory of the physical

[29] H.P. Stapp, *Mind, Matter and Quantum Mechanics,* Springer Verlag, Berlin 1993.

[30] *Idem*, *Mindful Universe: Quantum Mechanics and the Participating Observer*, Springer, Berlin 2007.

universe but the theory of experience with great emphasis on how it is felt by an observer. Stapp states clearly:

> In short, already the orthodox version of quantum mechanics, unlike classical mechanics, is not about a physical world detached from experiences; detached from minds. It is about predictions of relationships – entailed by a particular theoretical structure – between certain specified kinds of experiences.[31]

In his explanations, Stapp proceeds to combine the ontology of the potentialities suggested by Werner Heisenberg with *events* being the fundamental ontological quality demanded by the process ontology of Alfred North Whitehead. What turns out to be objective in Stapp's perspective are the potentialities for the events to occur and the events themselves. One of Stapp's intentions is to escape from the highly non-intuitive physicalist paradigm:

> My point here is that our most profound and deeply held intuition is not about the nature of the external physical world. It is rather that our human thoughts and efforts can make a difference in the behavior of our bodies. Our entire lives are based squarely on this perpetually re-validated intuition, as opposed to the proclamation of some philosophers, that our direct awareness of the physical efficacy of our thoughts is an illusion. The Heisenberg/Whitehead quantum ontology is thus concordant with both our most basic intuitions and with actual scientific practice. For this reason, I don't see why it should be difficult to shift science over to this improved way of conceptualizing nature and our role in nature.[32]

It must be rightly emphasized that Stapp is significantly shifting the epistemological focus from the supremacy of the external physical

[31] *Ibid.*, p. 103.
[32] *Ibid.*, p. 105.

Universe to the subjective experience and the priority of thought. The quantum event, according to Stapp, ties in itself the psychological and physical aspects of reality thereby enabling the intentionality of an observer to exert causal influence on the physical domain.

The next question which is vital to the entirety of Stapp's proposal is how to make the transition into the physical description of mental states. For that purpose, he suggests the notion of the template for action which he defines as:

> A *template for action* is defined to be a macroscopic (extending over a large portion of the brain) pattern of neurological activity that, if held in place for a sufficiently long period, will tend to produce a brain activity that will tend to produce an intended experienced feedback. This pattern of brain activity is the neural correlate (specified by a process 1 action) of a conscious effort to act in an intended way.[33]

In order to provide a link between the physical and psychological (mental), Henry Stapp resorts to the philosophy of William James. In particular, Stapp accentuates the role of *effort* to retain a certain mental state (such as perceiving an object or willing) as a specific element whereby the conscious experience is intensified. This is exactly the point where the ties with quantum mechanics come in. Repetitious efforts that generate conscious events will, by force of the quantum Zeno effect, stabilize the template of action and the corresponding pattern of neuronal activity stretching over large sections of the brain. In a nutshell, the quantum Zeno effect is the hindrance of the transitions between quantum states by measurements repeated with a sufficiently high rate. The measurement performed on a state that has not decayed yet causes the collapse to the undecayed state.[34] The

---

[33] *Ibid.*, p. 111.

[34] Cf. W.M. Itano, D.J. Heinzen, J.J. Bollinger, D.J. Wineland, *The Quantum Zeno Effect*, "Physical Review A" 1999, vol. 5 (41), pp. 2295–2300.

causal effect of the mental upon the physical is then explained by virtue of the prolonged effort exerted on the neuronal structures. Moreover, the decoherence process that would normally destroy the quantum superposition is perceived here as favourable. If decoherence acts to decompose the entangled state of the template of action, the Zeno effect occurring as a result of the applied effort prevents this decay thus stabilizing the neuronal structure responsible for a neuronal action under study. In order to address the question of why any neuronal structure would be in any kind of a quantum state, Henry Stapp simply postulates that each physical system – including the quantum states of the brain – participates in the quantum state of the Universe. Interestingly enough, Stapp distances himself from Eccles' ideas by stressing that the suggested 'mental force' acts not at the synaptical level of single neurons but on "an entire macroscopic neural pattern of brain activity".[35] As a summary of Stapp's reflections, the following quote seems particularly appropriate:

> Many scientists and philosophers have forced themselves to accept the rationally unsatisfactory and unsupported physicalist position in the mistaken belief that this is what basic physical theory demands. But the converse is true: contemporary physical theory demands certain interventions into the physical! The associated causal gap in a purely physically determined causation provides a natural opening to an interactive but non-Cartesian dualism.[36]

According to the following, an interesting pattern begins to emerge. A fundamental claim is being made that the physical deterministic Universe is not causally closed and the appropriate closure is effected by some – one could safely say – mystical intervention of the mental with the miraculous aid of quantum mechanics.

---

[35] H.P. Stapp, *Mindful Universe*, *op. cit.*, p. 114.

[36] *Ibid.*, p. 116.

5. David Chalmers – Fixing Quantum Mechanics

Similarly to Henry Stapp, in his efforts to apply quantum mechanics in the explanation David Chalmers places the greatest emphasis on the theory's central role in reconciling the experiential (psychical) real with the physical. A radical form of mind-body dualism comes to the fore. Moreover, Chalmers takes up an even a harder challenge by stating that an independently formulated theory of consciousness should be helpful in resolving the interpretational problems of quantum mechanics. To put matters bluntly, a philosopher tackles the interpretational problems of the theory without having any respectable background in theoretical physics. Be that as it may, the list of interpretations of quantum mechanics given in his seminal work, *The Conscious Mind*, is far from complete and does not seem to address most of the problems raised by theoretical physicists in this regard. It must be admitted, however, that Chalmers' proposal echoes the early excitement (and in some sense confusion) on where to draw the border between the mental and the physical in the effort to explain the non--intuitive ontology of the entangled states proposed by the theory's formalism. Chalmers' rightly notices the two components of quantum formalism proper to standard interpretations, namely the deterministic evolution of the wave function according to the Schrödinger equation and the measurement problem. His option is to move in the direction of eschewing the measurement problem by advocating Everett's interpretation, where all the possible outcomes of measurement can somehow coexist without having to be selected in the process of measurement. This is the moment, however, where the legitimate physics ends and philosophical speculation takes the lead.

The philosophical framework suggested by Chalmers is to recast Everett's interpretation of quantum mechanics in the context of his own theory of consciousness so that the interpretation's drawbacks can be eliminated. In regards to Everett's proposal, Chalmers strongly accents the fact according to this interpretation that the entire wave function bears an objective character that also extends to the mental realm. The reason why only a single experimental outcome is registered by each observer is that, following the measurement, each possible outcome is perceived by a different observer. This is supposed to reconcile the fact that the Universe is in an entangled state and yet each observer is aware of only a single experimental fact. Additionally, Chalmers warns against misinterpretations of Everett leading into the idea of the multiverse. His precise intention is as follows:

> The crucial element omitted from Everett's treatment is an analysis of the relationship between mind and body. Everett assumes that a superposed brain state will have a number of distinct subjects of experience associated with it, but he does nothing to justify this assumption. It is clear that this matter depends crucially on a theory of consciousness.[37]

The vital step taken now by Chalmers is to demonstrate how his own theory of consciousness completes Everett's interpretation of quantum mechanics. In particular, he claims that according to his theory it can be predicted that "a superposed brain state should be associated with a number of distinct subjects of discrete experience".[38] How does he accomplish that? In a nutshell, his theory of consciousness is aimed at explaining the so called *hard problem of consciousness*, that is, *qualia*. To be conscious, in Chalmers' terms, is to experience qualia. Consciousness is a fundamental reality in the Universe and cannot be explained by means of purely physical laws – one needs the set of psychophysical laws. The trickiest point to clarify is how the entangled state of a real physical system can be mirrored by the human mind. This is achieved by claiming "that consciousness arises from implementation of an appropriate computation".[39] By insisting that a computation of the entire entangled state can be decomposed

---

[37] D. Chalmers, *The Conscious Mind: In Search of a Fundamental Theory*, Oxford University Press, Oxford 1996, p. 324.
[38] *Ibid.*, p. 324.
[39] *Ibid.*, p. 326.

into its constituents thus preserving the Schrödinger evolution, Chalmers reaches its goal that each observer can experience only one experimental outcome and not the entire superposition simultaneously. He additionally supports his inference with his principle of the organizational invariance stating that physical and mental (psychophysical) structures exhibit the same organization. Thus Chalmers combines dualism with functionalism, albeit evidently in a non-reductive way. This allows him to proclaim the following final statement:

> it follows that a combination of the Schrödinger equation with an independently motivated theory of consciousness can predict our manifest image of the world. That is, the only physical principle needed in quantum mechanics is the Schördinger equation, and the measurement postulate and other basic principles are unnecessary baggage. To be sure, we need psychophysical principles as well, but we need those principles in any case, and it turns out that the principles that are plausible on independent grounds can do the requisite work here. This adds up to a powerful argument for taking the Everett interpretation seriously.[40]

As strange as it sounds, with the help of his philosophical stance Chalmers has solved the interpretational problem of quantum mechanics beyond any serious debate rooted in contemporary theoretical physics.

Conclusions

The detailed analysis of the different proposals in which quantum mechanics is employed in the explanation of mental phenomena reveals several interesting features. All of the cases discussed, except for that of Roger Penrose, give strong support to the mind-body dualism. It is particularly evident in case of John Eccles that quantum mechanics

[40] *Ibid.*, p. 326–327.

is sought to comply with an assumed dualist stance. The proposals of David Chalmers and Henry Stapp suggest that the "quantum support" for mind-body dualism draws upon the early disputes carried out by the founders of quantum mechanics on the role of the observer in a measurement performed on a quantum system. It is well known that the responses varied considerably (e.g., Bohr and von Neumann) leaving much freedom to draw the line between the quantum and classical regimes. Inasmuch as Bohr insisted that observation may be accomplished only in classical terms, Von Neumann claimed that quantum description may pertain to the macroscopic domain including the human mind.[41] In other words, it may be eventually the conscious act of an observer that decides on the outcome of an experiment.

Indeed, such solutions may be intuitively appealing but they offer no real explanation of the mind-body dualism. The collapse (reduction) of the wave vector, relegated to the action of the immaterial mind is currently well understood in terms of the process of decoherence being a purely natural phenomenon of cancelling the quantum superpositions upon their interaction with the microscopic environment. In short, quantum mechanics can no longer be considered as a mystical link filling the gap between psychophysical (mental) and physical laws. Moreover, both the dualist attitudes of Eccles, Stapp and Chalmers, as well as the allegedly physicalist stance of Roger Penrose and Stuart Hameroff, face the fundamental problem of how to correlate a particular quantum effect occurring in the brain tissue with a corresponding psychological function. This is most visible in the case of Henry Stapp, where a clearly psychological notion of the mental effort, which cannot be given any measurable scientific expression, is directly matched with the quantum Zeno effect. A similar problem arises between Chalmers' computational correlates of consciousness and Penrose and Hameroff's orchestrated objective reduction of the wave vector.

[41] J. von Neumann, *Mathematical Foundations of Quantum Mechanics*, Princeton University Press, Princeton 1955.

Besides the obvious objection raised to the possibility of the mental states being represented by entangled wave vectors strongly limited by the onset of decoherence, this idea faces further difficulties. It is not at all clear how the global quantum effects implied by the entanglement are supposed to have influence on the synaptic level and to regulate the rates of the neurotransmitter exchange. Even if this is to be accomplished by means of the objective reduction of the wave vector in the Penrose-Hameroff proposal, this is precisely the missing piece of physics that is yet to be discovered. This ties in directly with the second hope of the quantum mind theorists, namely, the theory's probabilistic nature allowing to relax the strict classical determinism thereby making room for uncaused behaviours proper, for instance, to the traditionally conceived notion of the free will. As indicated at the beginning, the probabilistic character of quantum mechanics is revealed only as a measurement is performed. The mental states represented by entangled wave vectors may evolve entirely deterministically without a trace of indeterminism unless an interaction with the environment occurs. This, in turn, invokes the fundamental problem of the interpretation of quantum mechanics and the physical reality that is attributed to the quantum states. By attempting to correlate a mental state with a wave vector, one immediately admits the reality of such a state. This comes to the fore especially in the case of the Penrose-Hameroff proposal but in a larger context of discussion it is a contentious issue.

The last issue that involves a legitimate scientific critique of the application of quantum mechanics to the explanation of the mental phenomena regards the more general framework of the cognitive research in which the cases studied are located. Two of them – that of Chalmers and Penrose/Hameroff – refer directly to the *computational paradigm* where the cognitive functions are modelled by algorithmic procedures. This may also involve quantum computation. Since computation is in its effect independent of physical substratum, the computational paradigm may of its very nature support dualist solutions to the mind-body problem. Interestingly enough, although Penrose and Hameroff regard their stance as a form of physicalism, their thesis on the non-algorithmic character of the activity of the mind leads them to speculate on the mind's intuitive access to the world of Platonic ideas. Although not strictly positioned on the mind-body axis, dualism seems to be an inescapable assumption. This problem may receive deeper clarification in the newer paradigm of cognitive science named the *embodied mind*, where the cognitive powers of the brain are determined by the functions of the body.[42] Moreover, the quantum solutions to the mind-body problem stipulate that the cognitive functions favour large (if not the entire) portions of the brain dictated by the non-local character of the corresponding wave function. This does not accord with the modular models of the brain put forward by the contemporary evolutionary psychology where the cognitive functions are of the adaptive origin.[43] If the evolution of the human species is to be considered as a part of the entire cosmic evolution following the initial singularity (the Big Bang), it is more proper to suppose that quantum mechanics (or more correctly quantum gravity) underpins the workings of the human mind. In short, it is much more likely that quantum mechanics (or more properly quantum gravity) explains the workings of the human mind rather than an a priori theory of mind would provide any clues as how to revamp quantum mechanics from its interpretational conundrum.

Inasmuch as the proposals put forward by Penrose/Hameroff and Eccles make an impression of being scientifically better justified, the other two – by Stapp and Chalmers – rely on philosophical speculation of high degree of arbitrariness. The amalgamate of philosophies utilized in a somewhat eclectic manner by Henry Stapp makes his proposal intuitively and rhetorically appealing but of no value from the point of view of its scientific explanatory power. Worse still, Stapp proposes a somewhat drastic interpretation of quantum mechanics in

[42] G. Lakoff, M. Johnson, *Philosophy In The Flesh: the Embodied Mind and Its Challenge to Western Thought,* Basic Books, New York 1999.

[43] J.H. Barkow, L. Cosmides, J. Tooby, eds., *The Adapted Mind: Evolutionary Psychology and the Generation of Culture*, Oxford University Press, Oxford 1992.

light of which the theory is not ultimately the theory of the physical Universe but the theory of conscious experience. In which case, science becomes an entirely subjective enterprise with pan-psychism as the primary ontology. In a similar vein, Chalmer's efforts to tackle the hard problem of consciousness end up in exploiting the physically unsupported interpretation of quantum mechanics by Everett, thereby grounding his solution of the hard problem of consciousness in a pan-psychic ontology. In short, quantum mechanics is trapped in a melange of philosophical (if not mystical) presuppositions where it serves to secure the causality implied by the laws of science within an a-priori assumed dualistic scheme. This damages not only the theory itself but neuroscience as a whole by forcing it to accept the reality of hypothetical entities which are foreign to its contemporary discourse.

Part II

At the Border Between Neuroscience and Philosophy

Łukasz Kurek
Jagiellonian University
Copernicus Center for Interdisciplinary Studies

Philosophical Issues in the Neuroscience of Mindreading
The Case of Belief[1]

Introduction

Mindreading is the ability to predict, explain, and describe people's behavior possessed by ordinary people and often used in social relations. It is an important area of research in cognitive and social neuroscience. Especially in the last decade, a great amount of empirical work related to the neuronal underpinnings of this ability has been made. What is interesting from a philosophical point of view is that various results of this research are connected with philosophical issues related to mindreading. Namely, these results support the claim that both of the leading philosophical theories of mindreading, the theory theory and simulation theory, are mistaken as to the nature of this ability. What is even more interesting is that there are some philosophical issues *in* the neuroscience of mindreading. In this paper, these philosophical issues shall be investigated using the concept of *belief*, a paradigmatic example of a concept involved in mindreading, as a case study. It will be claimed that the use of the concept of *belief* in neuroscientific theories rests on a philosophical assumption about the nature of belief.

---

[1] This paper was written within the research grant "The Limits of Scientific Explanation" sponsored by the John Templeton Foundation.

In the first part of the paper, two predominant theories explaining what belief is will be outlined, the so-called theory theory of mind and the simulation theory. In the second part of the paper, it will be argued that in the light of experimental data from neuroscience, both of these theories are inadequate in explaining what belief is. As a result, a more plausible account of the nature of mindreading and belief will be presented in brief. This hybrid account combines the insights of the theory theory and the simulation theory.

1. Theories of Mind and the Concept of *Belief*

Apart from the fact that the concept of *belief* is used in neuroscientific research, it is also an important issue in the philosophy of mind. One can trace the beginnings of the contemporary philosophical analysis of the concept of *belief* to the dawn of the analytical tradition in philosophical thought. What is more, the analysis of the perplexing issues connected with the concept of *belief*, such as opacity, the nature of content of belief, or the problem of internalism and externalism has been a sort of a showcase for analytical philosophers. Indeed, Bertrand Russell, who was one of the founding fathers of analytical philosophy, claimed that belief is "the central problem in the analysis of the mind".[2] According to Russell "believing seems the most 'mental' thing we do, the thing most remote from what is done by mere matter".[3]

[2] B. Russell, *The Analysis of Mind*, CreateSpace, Seattle 2011, p. 120.

[3] *Ibid.* It is important to note that for Russell and for many other philosophers in the analytical tradition, belief was a propositional attitude of a linguistic character. On this understanding, the psychological character of belief is deprived of any significance. What matters are the linguistic characteristics of the verb 'to believe'. The hidden assumption here is that a proper account of belief can be given by linguistic analysis of the reports of believing such as the sentence 'John believes it is raining'. In the contemporary philosophy of mind, however, the psychology of belief is of a greater importance. This is the reason why experimental work in the theory of mind is so important in philosophical account of belief.

Until the publication of the paper *Does the Chimpanzee Have a Theory of Mind* by psychologists David Premack and Guy Woodruf in 1978, mindreading had been an exclusively philosophical subject--matter.[4] However, in their seminal paper Premack and Woodruff not only introduced the concept of the *Theory of Mind* but also opened the field to empirical research. Shortly after that, the domain of mindreading became of great interest to cognitive scientists, and, in the last decade or so, to neuroscientists also. At present there is no doubt that a comprehensive account of mindreading should be of an interdisciplinary character because each of the disciplines involved in the explanation of mindreading (i.e. philosophy, psychology, and neuroscience) contributes significantly to this subject-matter.

It does not come as a surprise, therefore, that the two predominant theories which aim at explaining the ability to predict, explain, and describe people's behaviour are of a philosophical provenience. Both the theory theory and simulation theory provide not only philosophical accounts of how mindreading is possible but also form the conceptual basis for empirical research. Because of their general character they failed, however, to provide a set of criteria which would enable them to decisively settle the matter as to which of them is correct. Of course, there are some particular issues with which one theory deals better than the other. However, the definite resolution of the conflict between these theories seems to be very unlikely. What is important is that on both of these accounts of mindreading, beliefs play an important role. Namely, beliefs are necessary to perform at least the more complicated mindreading tasks. As it will become apparent, however, the role of belief is considerably more significant in the theory theory.

Apart from a few broad characteristics, one will not find, however, a precise definition of belief formulated in the theory theory or simulation theory. Concepts such as *belief*, located at the intersection of philosophy and empirical science, are better understood as being

---

[4] D. Premack, G. Woodruff, *Does the Chimpanzee Have a Theory of Mind*, "Behavioral and Brain Sciences" 1978, 1, pp. 515–526.

similar to law-cluster concepts, than to concepts of which one can provide a precise definition.[5] The notion of law-cluster concepts, introduced by Hilary Putnam, rests on the assumption that this type of concepts can be defined by laws in which they feature.[6] According to Putnam, the concept of *energy* is a good example of a law-cluster concept:

> It [the concept of *energy* – Ł.K.] enters into a great many laws. It plays a great many roles, and these laws and inference roles constitute its meaning collectively not individually. I want to suggest that most of the terms in highly developed science are law-cluster concepts, and that one should always be suspicious of the claim that a principle whose subject term is a law-cluster term is analytic. The reason it is difficult to have an analytical relationship among law-cluster concepts is that such a relationship would be one more law. But, in general, any one law can be abandoned without destroying the identity of the law-cluster concept involved, just as a man can be irrational from birth, or have a growth of feathers all over his body, without ceasing to be a man.[7]

On this account, to explain the meaning of a law-cluster concept one needs to point to the laws in which this concept features. This idea seems to be plausible in the case of at least certain natural sciences, for instance physics. As Putnam points out, there are many laws which involve the concept of *energy*. There are no differences in the case of other concepts used in physics. The situation is, however, distinct in the case of concepts related to mindreading, of which the concept of *belief* is a paradigmatic example. The difference lies in the fact that it is difficult to point to the laws (both psychological and neuroscientific) in which the concept of *belief* features. For that sim-

[5] J. Bermudez, *Philosophy of Psychology*, Routledge, London 2005, p. 10.

[6] H. Putnam, *Mind, Language, and Reality*, Cambridge University Press, Cambridge 1962, p. 52.

[7] *Ibid.*

ple reason, it is not possible to explain what belief is in the context of the appropriate scientific laws. Still, the concept of *belief* remains similar to the law-cluster concept in that it can be viewed as a theory-cluster concept.[8] The meaning of a theory-cluster concept can be reconstructed with a reference to different explanatory theories which feature this concept. In the case of the concept of *belief* these theories are of a philosophical, psychological and neuroscientific kind. Furthermore, instead of laws, the explanatory tools present in these theories include, for instance, indicating correlations between neuronal and psychological phenomena, or proposing psychological and neuronal mechanisms connected with attribution of belief or belief reasoning.

1.1 Theory Theory

There are different types of the theory theory accounts of mindreading.[9] The underlying assumption of them is that predicting, explaining, and describing people's behaviour is possible because agents possess a large body of information about mental states. This knowledge structure consists of rules, principles, or propositions which form a theory used by the agent in realizing mindreading tasks. This is why the discussed account of mindreading is called the theory theory

[8] J. Bermudez, *Philosophy...*, *op. cit.*, p. 10.

[9] The two leading types of theory theory are the child scientist theory (A. Gopnik, A. Meltzoff, P. Kuhl, *The Scientist in the Crib*, HarperCollins, New York 1999; J. Perner, *Understanding the Representational Mind*, MIT Press, Cambridge, Mass. 1991) and modularity theory (A. Leslie, *Pretending* and *Believing: Issues in the Theory of ToMM*, "Cognition" 1994, vol. 50, pp. 211–238; G. Botterill, P. Carruthers, *Philosophy of Psychology*, Cambridge University Press, Cambridge 1999). The main difference between these types of theory theory is that the proponents of the child scientist version claim that the acquisition of the mental concepts, rules, principles, and proposition is a matter of theory construction and revision, a process which analogous to the conduct of adult scientists. On the other hand, the proponents of the modularity theory claim that there is a Theory of Mind Module (ToMM) which utilizes the mental data which is to a large extent innate. However, the concept of belief seems to function similarly in both of these types of theory theory.

account. Of course, a large part of this knowledge structure cannot be accessed by the agent consciously, and in that part it is a tacit theory. On the other hand, some of the rules, principles, or propositions used in mindreading are accessible to the agent and can easily be made explicit. This happens when, for instance, the agent reasons about another person's behaviour.

The foundations of theory theory, in fact of the whole field of mindreading, were laid by Wilfrid Sellars who proposed that mental states can be viewed as theoretical states of a common sense psychology.[10] Common sense psychology was later often referred to, rather unfavourably, as 'folk psychology'. According to Sellars, mental states, such as belief or desires are just theoretical states which are used to predict, explain, and describe human behaviour. Furthermore, by claiming that there exist interconnections between mental states themselves the Sellarsian proposal was not only a first account of mindreading but also a first major opposition against the then prevailing philosophical behaviourism.

Not long after the publication of Sellar's influential book, *Empiricism and the Philosophy of Mind*, in 1955, a more elaborate account of mindreading was proposed by David Lewis. Lewis claimed that mental state terms are theoretical terms which can be defined in the context of the psychological platitudes which feature these terms:

> Think of commonsense psychology as a term-introducing scientific theory, though one invented long before there was any institution as professional science. Collect all the platitudes you can think of regarding the causal relations of mental states, sensory stimuli, and motor responses. Include only platitudes which are common knowledge among us—everyone knows them, everyone knows that everyone else knows them, and so on. For the meanings of our words are common knowledge, and I am going to claim

[10] W. Sellars, *Empiricism and the Philosophy of Mind*, (1955/2012) <http://www.ditext.com/sellars/epm.html>.

> that the names of mental states derive their meaning from these platitudes.[11]

Mental state terms form a collection of generalizations which are recognized by all mindreaders who live in the common cultural environment. These platitudes may be quite abstract or very specific, and they form a 'theory' which allows for the prediction, explanation and description of behaviour. What is crucial, however, is that it is assumed that there are causal relations between the entities denoted by observable and theoretical terms used in mentalizing.[12] One can differentiate between three types of such psychological laws: (1) laws relating observable inputs to mental states; (2) laws relating mental states to other mental states; (3) laws relating mental states to observable outputs (which are just types of behaviour). Here are some examples of psychological laws of the above-mentioned types connected with a mental state term 'belief':

(1) Persons who direct their perceptual attention to perceptible properties of events tend to believe that those events have those properties.
(2) Persons who believe that by performing action A they will achieve their goal B tend to form an intention to do A.
(3) Persons who believe that A, tend to assert A.

On this account, the meaning of mental state term 'belief' (i.e. the concept of *belief*) can be explained by the various different causal, or functional roles which this term performs. On the basis of the above--mentioned differentiation between types of psychological laws one can propose the following (admittedly very imprecise) definition of the term 'belief': (1) beliefs can be caused by certain observable

[11] D. Lewis, *Psychophysical and Theoretical Identifications*, [in:] *Readings in Philosophy of Psychology*, ed. Ned Block, vol. 1, Harvard University Press, Cambridge, Mass. 1980, p. 212.
[12] 'Mentalizing' is commonly used as a synonym for 'mindreading'.

inputs, such as perceptual attention; (2) beliefs can cause certain mental states, such as intentions; (3) beliefs can cause certain types of behaviour, such as assertions.

The idea that mental state terms are theoretical terms which feature in theory of mind appeared as natural to cognitive scientists, and was adopted by many researchers in this area. There are two reasons for this state of affairs. First of all, from the very beginning of the empirical research in cognitive science in the 1950's, the main aim for cognitive scientists was to explain various cognitive and behavioural capacities. One of the most striking examples of such a cognitive capacity is the capacity to predict, explain, and describe people's behaviour. After all, from a certain point of view, this ability is quite peculiar. On the one hand, it can be amazingly accurate. This supports the conclusion that there *really* are entities which mental state terms denote because the existence of such entities allows to explain the causal roles performed by those terms. On the other hand, mental states are invisible and there is no direct access to the mind of another agent. How is it possible, then, that the agent is able to mindread? Secondly, and more importantly, as it was pointed out by Stephen Stich, the dominant strategy of contemporary cognitive science is to posit an internally represented knowledge structure which is used to guide the execution of the capacity to be explained.[13] This strategy of explanation had been used in the case of many cognitive capacities, such as acquiring and using natural language, the ability to recognize visual objects, or the ability to make predictions and formulate explanations of the movement of middle-sized physical objects.[14] In the context of mindreading, the best analogue of using this explanatory strategy is, perhaps, the last of the mentioned cognitive abilities. According to the current body of

---

[13] S. Stich, *Deconstructing Mind*, Oxford University Press, Oxford 1996, p. 136.

[14] For such an account of the ability to acquire and use natural language see S. Pinker, *Learnability and Cognition*, MIT Press, Cambridge, Mass. 1989; in the case of recognition of visual objects see D. Marr, *Vision: A Computational Investigation into the Human Representation and Processing of Visual Information*, MIT Press, Cambridge, Mass. 2010.

knowledge about the ability to make predictions and formulate explanations of the movement of middle-sized physical objects, it is based on a common sense theory (naive physics).[15] Obviously, this theory is at odds with contemporary physics. One of the most striking findings in this domain is connected with the naive prediction of the motion of an object in a curved path. Experiments revealed that many people believe that "an object constrained to move in a curved path acquires a curvilinear impetus that causes it to follow a curved trajectory for some time after the constraints on its motion are removed".[16] This rule about the movement of middle-sized physical objects in a curved path can be viewed as a part of a common sense theory of motion.

The cognitive science version of the theory theory account of mindreading resembles the above-mentioned common sense theory of motion in that it is also a cognitive capacity commonly exercised by the agents which is to be explained with the help of experimental method. Furthermore, at least at the very beginning of the interest of cognitive scientists in the theory of mind, the strategy of explanation was to propose an internally represented knowledge structure which consists of observable terms and mental state terms and three types of relations between the types of terms described above. The philosophical assumption about the internal representation of the theory of mind is crucial in the context of neuroscience of mindreading. The best candidate of the place where this knowledge structure is represented is, after all, the brain of the agent. On this account is it claimed that there really *are* neural mechanisms which underpin the cognitive capacity to predict, explain, and describe people's behaviour in mental state terms such as 'belief'. This is the philosophical assumption which lies at the heart of the neuroscience of the theory of mind.

The field of the neuroscience of mindreading began to flourish when, with the use of functional neuroimagining, it was possible to

[15] S. Stich, *Deconstructing..., op. cit.*, p. 137.

[16] M. McCloskey, *Naive Theories of Motion*, [in:] *Mental Models*, eds. D. Gentner, A. Stevens, Lawrence Erlbaum, Hillsdale 1983, p. 311.

indicate which parts of the brain show increased activity when the agent performs mindreading tasks. However, one of the first obstacles for researchers working in this field was that mindreading can be an overwhelmingly complicated activity. There are many areas of the brain which are active when agents mindread. For instance, when the agent undertakes to explain the behaviour of another agent, this activity tends to consist in perceptual observation, language comprehension, working memory, and executive function, to name but a few of the cognitive processes involved. The search for the neural bases of these cognitive processes is fascinating in itself, they do not seem to be, however, particular to mindreading. To commence with the neuroscience of mindreading one needs to, first of all, identify processes which are especially important for mentalizing. A proposal of such an identification has been made by Rebecca Saxe and Liane Young, neuroscientists working in the field of theory of mind. They identified three cognitive processes which "appear to be disproportionately necessary for reading and understanding other people and what they are thinking" (in the parentheses the brain regions which are associated with those processes are identified):

(1) Process responsible for representing people and social relations (e.g. medial prefrontal cortex, MPFC).
(2) Process responsible for representing representations (e.g. left temporo-parietal junction, LTPJ).
(3) Process responsible for representing *mental* representations (e.g. right temporo-parietal junction, RTPJ).[17]

The three cognitive processes which are 'disproportionately necessary' for mindreading can be identified both on the conceptual level and at the empirical level. At the conceptual level, one can point to the

---

[17] R. Saxe, L. Young, *Theory of Mind: How Brains Think about Thought*, http://moralitylab.bc.edu/publications/, pp. 3–4 [forthcoming in K. Oschner, S. Kosslyn, *The Handbook of Cognitive Neuroscience*, Oxford University Press].

fact that, first of all, mindreaders have to differentiate between situations which involve predicting, explaining, and describing the behaviour of unintentional physical objects (when it is not necessary to use mental state terms) and situations when one undertakes to predict, explain and describe behaviour of other agents (when mental state terms are commonly used). Secondly, at the further stage of mindreading, it is necessary to *represent* other agent's representations (this is an ability to understand that other agents may have a different perspective on the same events). It includes, for instance, an ability to understand that agents have a different representation of an object when they look at it from a different perspective. Thirdly, the last cognitive process involved in mindreading is devoted exclusively to thinking about other people's mental states, such as beliefs. What is interesting from the neuroscientific perspective, it was possible to support this conceptual distinction with empirical research.

As for the neural identification of the cognitive processes responsible for representing people and social relations, Saxe and Young associate it with the activity of the MPFC.[18] Experiments reveal that there is no high level activity in this brain region when agent reasons on representations, or interactions between physical objects. Only tasks which involve thinking about people or social interactions appear to heavily involve MPFC. Saxe and Young suggest, therefore, that the activity in MPFC is particular to social cognition.

The second of the cognitive processes related to theory of mind, the process responsible for representing representations, can be associated with the activity of LTPJ. Saxe and Young suggest that a high activity of LTPJ can be observed in tasks which require reasoning about different types of representations, not necessarily mental representations.[19] These reasoning tasks include, for instance, physical representations about the world such as pictures. Furthermore, subjects with lesions of the LTPJ have difficulties in reasoning about different

[18] *Ibid.*, p. 7.

[19] *Ibid.*, p. 6.

types of representations, both mental and physical. What is also interesting is that these subjects do not have problems with higher-level reasoning which does not include representations (i.e. naming the ink colour of the word 'green' printed in red ink). This supports the differentiation between the left dorsal lateral prefrontal cortex (LDLPFC) which is associated with high-level reasoning, regardless of whether there is a representational content included and LTPJ (associated with only one type of reasoning, namely 'representational reasoning').

The third cognitive process connected with mindreading, and perhaps the most interesting one, is the process responsible for representing *mental* representations. Saxe and collaborators associate it with the activity within RTPJ.[20] This brain region shows increased activity in reasoning which involves mental states such as beliefs; it is not active in the case of other types of high-level reasoning, including 'representational reasoning' which does not involve mental representations, and even reasoning about other agents and social relationships which does not involve mental states. Despite the fact that MPFC and RTPJ are usually activated together (which does come as a surprise; thinking about other people and social relationships often involves thinking about mental states such as beliefs), one can point to a much revealing dissimilarity between the activity of those two brain regions. Saxe and collaborators claim that activity within RTPJ is high when experimental tasks include thinking about true and false beliefs, but it is considerably lower if it involves thinking about the physical traits of an agent, her personality traits, or preferences.[21] On the other hand, one can observe high activity in MPFC in all of the above--mentioned tasks. What is even more interesting, a temporal 'switching-off' of the RTPJ with the use of transcranial magnetic simula-

---

[20] R. Saxe, N. Kanwisher, *People Thinking About Thinking People. The Role of the Temporo-Parietal Junction in "Theory of Mind"*, "Neuroimage", 2003, vol. 19, pp. 1835–1842; R. Saxe, L. Powell, *It's the Thought That Counts: Specific Brain Regions For One Component of the Theory of Mind*, "Psychological Science", 2006, vol. 17, pp. 692–699.

[21] R. Saxe, L. Young, *Theory of Mind..., op. cit.*, p. 8.

tion produced further evidence of the notable role of this brain region in mindreading.[22] In the tasks involving moral judgments, this kind of a controlled 'lesion' changes the agent's decision. After the 'lesion', the beliefs and other mental states of the agent become of little relevance to the outcome of the moral judgment. The unintended actions of the agents become more blameworthy and intended actions become less blameworthy. These results support the claim that RPTJ is particularly involved in the cognitive process responsible for portraying mental representations.

One needs to stress, however, that the above-mentioned description of the neuronal basis of mindreading is, obviously, to a large extent incomplete. For instance, Saxe and Young point to the important role in the theory of mind of a different brain region, the already mentioned LDLPFC. The activity of this brain region is not particular to this cognitive capacity. To describe its function most generally, LDLPFC "shows a high response to stimuli requiring difficult reasoning, especially for balancing competing ideas or responses".[23] In the case of the theory of mind, however, damage to this brain region makes it difficult to resolve cognitive conflict in the case of differences in the representations between the agent who performs a mindreading task and the agent who is the target of mindreading. Experimental work revealed that patients with LDLPFC lesions have difficulties with resolving the conflict between their beliefs and the beliefs of the targets of their mindreading. In such cases, these subjects favour using their own beliefs in the explanation of the behaviour of the other agent.[24] This cognitive bias is, of course, absent in a situation where there is no conflict between the beliefs of the mindreader and his target. Subject with lesions in the LDLPFC do not have

[22] L. Young, J. Camprodon, M. Hauser, A. Pascual-Leone, R. Saxe, *Disruption of the Right temporo-parietal Junction with Transcranial Magnetic Simulation Reduces the Role of Belief in Moral Judgements*, "Proceedings of the National Academy of Sciences" 2010, vol. 107, pp. 6753–6758.

[23] *Ibid.*, p. 5.

[24] *Ibid.*

difficulties with thinking about beliefs in those situations.[25] This brain region is one of the many parts of the brain which subserve the cognitive process responsible for representing people and social relations.

To summarize the theory theory account of mindreading, the assumption that agents possess an internalized knowledge structure which enables them to predict, explain, and describe people's behaviour in mental state terms is supported from different areas of both science and philosophy. In this paper, the focus of analysis is on the neuroscientific evidence which supports the theory theory. This understanding is based on an assumption that belief is individuated by the functional role that it plays in generating behaviour and other mental states. Furthermore, beliefs are understood to be represented in the brain of the agents. Neuroscientific research points to the conclusion that there are neural mechanisms which are subservient mental state attributions of mental states such as belief. On the basis of the empirical research mentioned, a good place where one can look for such mechanisms is RTPJ.

1.2 Simulation Theory

Until the late 1980's, the theory theory account of mindreading was 'the only game in town'. The reasons for this state of affairs have already been mentioned: mindreading was, perhaps, one of the most interesting examples of cognitive capacities to be explained and the theory theory explanation is based on the dominant explanatory strategy in cognitive science. However, in 1986 in a seminal paper *Folk Psychology as Simulation*, Robert Gordon proposed a radically new account of mindreading, well supported both from a theoretical and empirical viewpoint.[26] Gordon's primary claim was that there is no in-

---

[25] *Ibid.*

[26] Robert Gordon, *Folk Psychology as Simulation*, "Mind and Language" 1986, vol. 1, pp. 158–171.

ternally represented knowledge structure which enables mindreaders to predict, explain, and describe people's behaviour. Instead of using this knowledge structure (which enables, for instance, the ascription of a certain belief to the target and the use of this belief in reasoning about other mental states and behaviour of the target), mindreaders imagine how the world is from their perspective. In this kind of simulation, there is no need for mental state terms. Agents who are simulators think about the world from the perspective of other agents but without reference to their psychological states.[27]

There is also a different account of simulation which has proven to be more influential. According to this account of simulation, it is not the case that in simulating the mind of another agent mental state terms such as belief are irrelevant.[28] On the contrary, their role is quite important. However, according to Alvin Goldman, in the case of simulation theory there is no need to refer to the internally represented knowledge related to mental states such as psychological laws which are supposed to bind them. Instead of using such generalizations, we use our own decision-making process to run a simulation of what we

[27] It is perhaps interesting to note that the beginnings of simulation theory in the history of thought are much more venerable. Two thinkers can be viewed as philosophical precursors to this account of mindreading: David Hume and Adam Smith. Obviously, both Hume and Smith did not use the term 'simulation' to refer to the ability to predict, explain, and describe people's behavior. The term which is crucial in their thought related to this topic is 'sympathy' (which can be understood as today's 'empathy'). Here is the relevant passage from Hume: "When I see the *effects* of passion in the voice and gesture of any person, my mind immediately passes from these effects to their causes, and forms such a lively idea of the passion as is presently converted in the passion itself" (D. Hume, *A Treatise of Human Nature*, 1739/2010, <http://www.gutenberg.org/files/4705/4705-h/4705-h.htm>). And an equally interesting quotation from Smith: "When we have read a book or poem so often that we can no longer find any amusement in reading it by ourselves, we can still take pleasure in reading it to a companion. To him it has all the graces of novelty; we enter into the surprise and admiration which it naturally excites in him, but which it is no longer capable of exciting in us; we consider all the ideas which it presents rather in the light in which they appear to him, than in that in which they appear to ourselves, and we are amused by sympathy with his amusement which thus enlivens our own" (A. Smith, *The Theory of Moral Sentiments*, <http://ebooks.gutenberg.us/WorldeBookLibrary.com/moralsent.htm>, 1759/2008).

[28] A. Goldman, *Simulating Minds: The Philosophy, Psychology, and Neuroscience of Mindreading*, Oxford University Press, Oxford 2006, pp. 28–29.

would do in a scenario where we have the same beliefs and desires as the target of mindreading. In other words, in simulation the decision-making process of the simulator is running off-line so that the decision which is generated in this mode in not acted upon but, instead, attributed to the target.

The popularity of this account of simulation over the more radical one, which is devoid of mental state terms, can explained by pointing to the difficulties connected with proposing a detailed explanation of how it is possible to imagine the world from the perspective of another agent without reference to mental states. Some simulationists, like Jane Heal, have tried to shed some light on this problem (Heal uses the term 'replication' instead of 'simulation'; the idea appears to be, however, very similar):

> On the replicating view psychological understanding works like this. I can think about the world. I do so in the interests of taking my own decisions and forming my own opinions. The future is complex and unclear. In order to deal with it I need to, and can, envisage possible but perhaps non-actual states of affairs. I can imagine how my tastes, aims, and opinions might change, and work out what would be sensible to do or believe in the circumstances. My ability to do these things makes possible a certain sort of understanding of other people. I can harness all my complex theoretical knowledge about the world and my ability to imagine to yield an insight into other people without any further elaborate theorizing about them. Only one simple assumption is needed: that they are like me in being thinkers, that they possess the same fundamental cognitive capacities and propensities as I do.[29]

The difficulty with this account of mindreading is that, despite the fact that there is obviously no explicit attribution of beliefs and other

[29] J. Heal, *Replication and Functionalism*, [in:] J. Butterfield *Language, Mind and Logic*, Cambridge University Press, Cambridge 1986, p. 47.

mental states to other people, the conceptual framework of the theory of mind seems to be necessary to *simulate* how the mindreader would behave if put in the situation of the target person. Heal clearly refers to simulation as working out what would be sensible to do or *believe* in the circumstances of another agent. This account presupposes then, that the ability to predict, explain, and describe people's behaviour features mental state terms. Goldman proposes an explanation of the aversion of radical simulationists towards the more standard account of simulation which can be discussed on the context of empirical science. He claims that advocates of this version of mindreading appear to treat simulation theory as an a priori thesis based on conceptual considerations, and not a scientific hypothesis which connects mindreading with cognitive processes such as decision making, or with their neural correlates.[30]

There is no doubt, however, that the scope of the *internal knowledge structure* which forms the basis of mindreading on the theory theory account is severely limited in both versions of simulation theory. This is especially clear when one tries to compare the role of belief in both of these accounts. To recall, the understanding of the concept of *belief* which is assumed in this paper, *belief* is a theory-cluster concept which can be defined by indicating its role in various explanatory theories (philosophical, psychological, and neuroscientific) in which it features. On the theory theory account, belief features heavily in these theories, and it is often used in explaining the ability to mindread others. In the simulation theory, it is argued that there is *no body of information* which features the concept of *belief*. Even more so, possession of such a large body of information concerning mental states and relations between these states and between these states and behaviour is a highly implausible assumption. This seems to a certain extent implausible, as it is difficult to propose a coherent account of mindreading which could do completely without the agent's assumption of some body of information, most of it implicit, about men-

[30] A. Goldman, *Simulating Minds...*, *op. cit.*, p. 20.

tal states. For instance, often to obtain the pretend inputs for simulation, agents need to form representations of other agent's thoughts, and how could they do so without possessing at least some knowledge about these mental phenomena? Despite this, it is evident that, according to simulation theory, the role of belief (and other mental states) is much more limited than on the theory theory account.

The main difference between these two stands, therefore, boils down to different accounts of the scope of the internal knowledge about mental states that the mindreaders possess, and to different mechanisms proposed to explain mindreading. As a consequence, they propose different roles for the same mental states. In simulation theory, what is enough to mindread is that agents themselves possess decision-making capacities which can be run offline. This requires that the input in the process of simulation should be a pretend one. On the basis of such a pretend input, and the functioning of the agent's decision-making process run offline, a pretend output (the decision of the target agent) is generated. A much revealing description of the simulation theory is that "mindreaders use their own minds to *mirror* or *mimic* the minds of others".[31]

Many readers will associate talk of 'mirroring' the minds of others with the discovery of mirror neurons. The fact that both philosophers who work on mindreading and neuroscientists use the term 'mirroring' is no coincidence. In fact, one of the first philosophers who made use of the mirror-neuron system in their argumentation was Alvin Goldman in his support of simulation theory. Together with Vittorio Gallese, one of the discoverers of mirror neurons, he wrote a paper in which the mirror neuron system was proposed to be a mechanism of primitive, simulation-like mindreading.[32] Indeed, the discovery of the mirror neuron system proved to be a powerful argument for the simulationists.

---

[31] *Ibid.*

[32] V. Gallese, A. Goldman, *Mirror Neurons and the Simulation Theory of Mindreading*, "Trends in Cognitive Sciences", vol. 2, pp. 493–501.

Mirror neurons for action were first discovered in the ventral premotor cortex (area F5) of macaque monkeys, using the technique of single cell recording.[33] The extraordinariness of mirror neurons consists in the fact that they discharge both when the acting monkey executes a goal-directed action, and also when an observing monkey watches the same type of action performed by someone else.[34] Before the discovery of mirror neurons, the premotor cortex was usually associated with action planning so the activity within this brain region in response to the mere observation of the action of others was a surprise. Research on mirror neurons also revealed that the activation of mirror neurons in the observer in response to the observed action had the same functional properties as performing the same type of action by the observer. The only difference is that the instruction to act generated by mere observation of the act is inhibited. For instance, the experiment of Fadiga and collaborators revealed that in humans the same muscle group are facilitated in the observer as in the acting agent.[35] In the observer, however, the muscular activity is inhibited.

From the empirical standpoint, the mirroring attribution of mental states is especially well supported in the case of emotions and intentions.[36] Arguably, attributing emotions, and intentions are examples of mindreading, just as attributing belief is. Attributing emotions and intentions are examples of cognitive capacities, the results of which can be used to explain, predict, or describe behaviour of other agents. Let us focus on attributing emotions, as the research on this topic is more extensive and seems to support the simulationist account even

---

[33] G. Rizzolatti, L. Fadiga, V. Gallese, L. Fogassi, *Premotor Cortex and the Recognition of Motor Actions*, "Cognitive Brain Research" 1996, vol. 3, pp. 131–141; V. Gallese, L. Fadiga, L. Fogassi, G. Rizzolatti, *Action Recognition in the Premotor Cortex*, "Brain" 1996, vol. 119, pp. 593–609.

[34] V. Gallese, A. Goldman, *Mirror Neurons...*, *op. cit.*, p. 495.

[35] L. Fadiga, L. Fogassi, G. Pavesi, G. Rizzolatti, *Motor Faciliation During Action Observation: A Magnetic Stimulation Study*, "Journal of Neurophysiology", vol. 73, pp. 2608–2611.

[36] In the case of intention see, for instance, M. Iacobini, I. Molnar-Szakacs, V. Gallese, G. Buccino, J. Mazziotta, G. Rizollatti, *Grasping the Intentions of Others with One's Own Mirror Neuron System*, "PLoS Biology", 2005, vol. 3, pp. 529–535.

better. What is more, this is a more 'primitive' type of mindreading, because it is comparatively simple, primitive, automatic, and often below the level of consciousness.[37] Attributing intentions seems to be a more complicated matter because, at least at the conceptual level, it is connected with goal related action, and, as such, requires a lot more from the agent than 'mere' emotion attribution.

The most interesting research in this area is related with face-based emotion recognition (FaBER). FaBER consists only in recognizing emotion types (usually the basic emotions of fear, anger, disgust, happiness, sadness, and surprise).[38] It does not include recognizing the content of the emotion (for instance, at whom a target person is angry, and why). Furthermore, the regions in the brain associated with emotional behaviour are evolutionarily 'older' than regions associated with explicit reasoning. These insights support the claim that there is no *inherent knowledge structure*, whether it is explicit or implicit, which subserves FaBER. Yet, agents are amazingly proficient at attributing emotions based on facial expressions.[39]

The neuroscientific evidence supporting simulation theory should reveal, therefore, that there is a correlation between the deficit in experiencing a certain type of emotion and the deficit in recognizing this type of emotion. On this account, the impairment of the ability to experience a certain type of emotion should cause the impairment of the ability to simulate it by the agent and, as a consequence, it would lead to difficulties with attributing this emotion. As it was revealed by experimental work, this type of correlation is present in the case of several types of emotions, including fear, disgust, anger, or the 'non-basic' emotion of guilt.[40] Fear and disgust are the most clear-cut examples.

---

[37] A. Goldman, *Simulating Minds…*, *op. cit.*, p. 113.

[38] *Ibid.*

[39] Paul Ekman, Wallace Friesen, *Unmasking the Face: A Guide to Recognizing Emotions From Facial Expressions*, Malor Books, Cambridge, Mass. 2003.

[40] R. Adolphs, H. Damasio, D. Tranel, G. Cooper, A. Damasio, *A Role for the Somatosensory Cortices in the Visual Recognition of Emotion As Revealed by Three-Dimensional Lesion Mapping*, "Journal of Neuroscience" 2000, vol. 20, pp. 2683–2690; R. Adolphs, *Fear and the Human Amygdala*, "Journal of Neuroscience" 1995, vol. 15, pp. 5879–5891.

The brain region which is strongly associated with the emotion of fear is the amygdala. Ralph Adolphs and his collaborators investigated whether damage to this region of the brain influences the face based recognition of fear. One of their patients, SM, suffered from a bilateral destruction of her amygdalae. This is how Antonio Damasio, a co-investigator in this case, describes SM 'experience' of fear:

> SM does not experience fear in the same way you or I would in a situation that would normally induce it. At a purely intellectual level she knows what fear is supposed to be, what should cause it, and even what one may do in situations of fear, but little or none of that intellectual baggage, so to speak, is of any use to her in the real world. The fearlessness of her nature, which is the result of the bilateral damage to her amygdalae, has prevented her from learning, throughout her young life, the significance of the unpleasant situations that all of us have lived through.[41]

SM was also tested on FaBER tasks.[42] In these tasks, subjects are presented with photographs or videos that depict facial expressions, and are asked to identify emotions which they associate with these expressions. The results of the experiments revealed that her association of the emotion of fear with the facial expression of fear is weak. Her results for the association of fear with its facial expression correlated considerably less with normal ratings than the results of the 12 control subjects with lesions to different parts of the brain.

A similar example of the correlation between being able to experience a certain type of emotion and being able to recognize it can be found in the case of disgust. The brain region which is associated with experiencing disgust is the anterior insula region.[43] Using fMRI

---

[41] A. Damasio, *The Feeling of What Happens*, Harcourt Brace, New York 1999, p. 66.

[42] R. Adolphs, D. Tranel, H. Damasio, A. Damasio, *Impaired Recognition of Emotion in Facial Expressions Following Bilateral Damage to the Amygdala*, "Nature", 1994, vol. 372, pp. 669–672.

[43] L. Phillips et al., *A Specific Neural Substrate For Perceiving Facial Expressions of Disgust*, "Nature" 1997, vol. 389, pp. 495–498.

techniques, Lynnette Phillips and her collaborators were able to establish that when subjects observed facial expressions of disgust, the right insula was activated. The conclusion of the experimenters was that 'appreciation of visual stimuli depicting others' disgust is closely linked to the perception [i.e., experience] of unpleasant tastes and smells'.[44] Contrary to the above-mentioned experimental research on fear, this study points to a correlation between the experience of disgust and disgust recognition in healthy subjects.

To summarize the considerations connecting simulation theory and FaBER, the experimental work in this area strongly favours this theory over the theory theory account. Based on this research, it is implausible to assume that there is an implicit knowledge structure which associates beliefs about different types of emotion with beliefs about their behavioural features and guides the agent in emotion recognition. What is more, often the subjects with lesions to the brain regions associated with different types of emotions, still can provide explicit information about the type of emotion they have difficulties recognizing, such as information about situations which typically elicit this type of emotion. A subject who preserved such knowledge was, for instance, SM.[45]

2. Two Types of Mindreading Processes and the Nature of Belief

As the above neuroscientific evidence suggests, there are strong arguments in favour of both the theory theory account of mindreading and the simulation theory. The picture which emerges from the above analysis of the experimental work reveals that perhaps both of these accounts are plausible, but in different domains of mindreading. The theory theory account appears to deal better with the attribution

---

[44] *Ibid.*, p. 496.

[45] R. Adolphs et al., *Impaired Recognition...*, *op. cit.*, p. 670.

and recognition of more complex mental states, especially beliefs. In the case of the attribution of belief, complex cognitive processes are used which require access to a large body of information. On the other hand, more 'primitive' instances of mindreading, such as emotion recognition and attribution, do not seem to require such information. The process of the attributing and recognition of these mental states is fast, automatic and rather unreflective. Insights from various disciplines investigating cognition are coherent with this proposal.[46]

2.1 Dual-Process Models of Social Cognition

The distinction between simple and complex cognitive processes has been proposed in various domains even before the development of the neuroscience of mindreading.[47] In the context of social neuroscience, a comprehensive review of the features of the two types of processes was made by Matthew Lieberman.[48] Although this is a review of the research within the domain of social neuroscience, not within the domain of the neuroscience of mindreading, the subject matter of these two domains is similar, and the apparent differences only terminological. Namely, according to Lieberman, social neuroscience pertains to representing psychological states and traits of other agents, or experiencing mental states of others (such as emotions).[49] These matters are crucial for mindreading.

The two mentioned processes which can be distinguished on the basis of the current state of the art in social neuroscience are the reflexive

---

[46] A particularly interesting example is the domain of number cognition. This ability is explained by proposing two processes: an efficient one and a flexible one. The first one is used to process small numerosities, up to three or four elements. The second is used to process larger numerosities. For an overview of the research in this domain and an example of its use in philosophical considerations see Bartosz Brożek, *The Return of Empirical Subject in the Philosophy of Mathematics*, [in:] *Between Philosophy and Science*, Copernicus Center Press, forthcoming.

[47] See, for instance, S. Chaiken, Y. Trope, *Dual-Process Theories in Social Psychology*, The Guilford Press, New York 1999.

[48] Matthew Lieberman, *Social Cognitive Neuroscience: A Review of Core Processes*, "Annual Review of Psychology" 2007, vol. 58, pp. 259–289.

[49] *Ibid.*, pp. 263–265.

system and the reflective system.[50] Similarly to the above considerations concerning neuroscience of mindreading, these systems can be distinguished, first of all, at the functional level. Some functional features of the reflexive system include: fast operation, slow learning, spontaneous responses, nonreflective consciousness, being phylogenetically older than the reflective system, typically sensory, and representation of common causes.[51] The functional features of reflective system, on the other hand, include: slow operation, fast learning, reflective consciousness, intentional processes, typically linguistic, being phylogenetically newer than the reflexive system, representation of special cases, and abstract concepts.[52] The features of the reflective and reflexive system correspond to the characteristic of cognitive processes responsible for mindreading proposed, respectively, by the theory theory and simulation theory. On the theory theory understanding of mindreading, the required cognitive process should be flexible, complex, and using a large body of information. Simulation theory proposes, however, that these cognitive processes are inflexible, simple, and do not require additional information to operate.

Secondly, the reflective and reflexive systems can be distinguished at the level of the brain regions which are associated with their functioning. Brain regions associated with reflective system include, *inter alia*, lateral prefrontal cortex (LPFC), medial prefrontal cortex (MPFC), and lateral parietal cortex (LPC).[53] Brain regions associated with the reflexive system include, for instance, the amygdala, basal ganglia, and ventromedial prefrontal cortex (VMPFC).[54] Also in this case there is a clear correspondence between brain regions associated with these processes and, respectively, brain regions associated with the theory theory of mindreading and simulation theory.

---

[50] *Ibid.*, p. 261.
[51] *Ibid.*, p. 261.
[52] *Ibid.*
[53] *Ibid.*
[54] *Ibid.*

2.2 The Hybrid Theory of Mindreading

Empirical research within social psychology, cognitive science, neuroscience, and other disciplines related with mindreading have caused a significant change in the philosophical debate between the proponents of theory theory and simulation theory. Until very recently, it was considered that one of these accounts is incorrect because they cannot be reconciled and no other coherent explanation of mindreading is possible. However, a substantial amount of experimental data from various disciplines supports the claim that, indeed, both of these accounts are in part correct. This is the reason why in recent years hybrid accounts of mindreading have become popular and developed in detail by influential thinkers in the field.[55] These accounts often make use of the insights from the theory theory in relation to high-level mindreading (related with such mental states as beliefs), and the insights from simulation theory are used in relation to low-level mindreading (related with, for instance, emotions).

It is interesting to consider the role of belief in the light of the above considerations. Two interesting conclusions can be drawn here. First of all, the scope of predicting, explaining and describing other people's behaviour which requires the use of the concept of *belief* is narrower than proponents of theory theory argue. Various, and important types of mental states – for instance emotions, or intentions – are attributed without the need to form any beliefs about the agents, which are based on common sense psychological theory. These states are attributed by a process of automatic, fast operating simulation processes.

Secondly, it is not the case, as certain simulationists argue, that mindreading abilities can be explained without reference to the concept of *belief* (in case of radical simulationists), or that this concept can be explained by simulational processes alone (in the case of more

---

[55] These include, for instance, A. Goldman, *Simulating Minds..., op. cit.*, or S. Nichols, S. Stich, *Mindreading: An Integrated Account of Pretence, Self-Awareness, and Understanding of Other Minds*, Oxford University Press, Oxford 2003.

standard accounts of simulation). On the contrary, in high-level mindreading, which includes thinking about other people's thoughts, this concept functions as an important element of the explanatory strategy and it is connected with a theory-like body of information. Admittedly, it is difficult to provide a precise definition of this concept. Some of the above mentioned features of belief, such as its involvement in flexible, conscious reasoning, and its grounding in the body of information which includes relations between belief and other mental states and relations between belief and behaviour do, however, explain the nature of belief to some extent.[56]

Conclusion

In the first part of the paper, two leading theories aiming at the explanation of the ability to predict, explain, and describe people's behaviour were presented: theory theory and simulation theory. It was claimed that both of these theories are partially supported by empirical research conducted in cognitive and social neuroscience. Furthermore, it was argued that on the theory theory account, the role for belief is much more significant for explaining the ability to mindread others than in the simulation theory. In the second part of the paper, it was proposed that the two theories can be partly reconciled by introducing a hybrid theory which links important insights coming from both of these accounts. As a consequence, on this account the nature of belief is different from the nature of belief proposed in both of these stands towards mindreading. Belief is neither an all-encompassing element of this ability, nor an element that a theory of mindreading can do without. Rather it is a mental state which features only in high-level mindreading.

[56] More on belief, in the context of the issue of self-deception and from the perspective of both philosophy and cognitive science, can be found in M. Hohol, P. Urbańczyk, *Self-deception. Between Philosophy and Cognitive Science*, this volume.

Julia Stanek
Jagiellonian University

Self-Consciousness
From Philosophy to Neuroscience[1]

Introduction

The issues of consciousness and self-consciousness have been the subject of great interest since ancient times. How can we understand the phenomenon of self-consciousness? Where can we find its source? In what way does something like consciousness and self-consciousness emerge from physical events in our brains? Scientists in different fields have been investigating these and many other questions related to the self. When considering the problem of what self--consciousness is, it seems, at least at first glance, that the answer has to be exceptionally simple. As a person is self-conscious, she just knows that she is in this "state", the existence of it is evident and inseparably connected with her own being. Researchers have immense difficulty in explaining how we can define and understand self-consciousness and the self. Some scientists believe that the self can be identified with memory or memories, the others focus only on the topic of the ability to distinguish ourselves from external world and unity of the self, while, among neuroscientists and psychologists, consciousness is often identified with attention. In this paper, the expression self-

---

[1] This paper was written within the research grant "The Limits of Scientific Explanation" sponsored by the John Templeton Foundation.

-consciousness stands for the permanently emerging experience of being distinguished from the surrounding environment and the experience of being the creator of our thoughts and actions. Here the term "self" is used in the sense of the bearer of self-conscious states.[2] Additionally, with regard to the fact that the issue of being a person is immanently connected with matter of self-consciousness and consciousness, in this paper the concept of person is identified with self.

The problem of the self is extremely complex and has many aspects. It appears that only an interdisciplinary approach is able to reveal its nature. Recently the development of non-invasive methods for investigating the structure and function of the brain (such as functional brain imaging) has significantly contributed to the growth of research possibilities in the questions associated with consciousness and self-consciousness. Further in this paper the findings of researches from various scientific fields concerning the essence of the self are discussed. In particular, I concentrate on the studies of self-conscious states from the neuroscience perspective. Recently, neuroscience has been in the limelight and it is believed that research in this discipline is close to exploring the nature of the self. Some neuroscientists have even postulated that it is the next and last step in gaining full knowledge about the nature of self.[3] Obviously, it is still debatable whether it is indeed reachable. In this paper I try to present in what way it may be accessible. Also here the current contributions of neuroscience to self-consciousness studies are discussed. One of the main goals of this paper is to clarify whether the methods of investigation and basic philosophical foundations of neuroscience are appropriate for understanding the problem of the self. Finally, I comment on the methods and foundations of neuroscience.

---

[2] This definition is also used by Albert Newen and Anika Fiebich in the paper: *A Developmental Theory of Self-models: Individual-cognitive and Social-cognitive Dimensions of Self-consciousness,* [in:] *Social Roots of Self-Consciousness*, eds. W. Mack, G. Reuter, Akademie Verlag, Berlin 2009, pp. 161–186.

[3] For example: V. Lamme in: *Can Neuroscience Reveal the True Nature of Consciousness?,* available from <http://www.nyu.edu/gsas/dept/philo/courses/consciousness05/LammeNeuroscience.pdf>.

There is a wide range of different approaches that have been used to study self-consciousness. Each of these approaches focuses on diverse aspects of the self, using different methods. Not surprisingly that in consequence, they reach various conclusions about the nature of the self. In general, studies on self-consciousness are conducted in the following research areas:

- philosophical (above all phenomenology and the philosophy of mind);
- psychological and sociological;
- clinical (mostly neurology, psychopathology and psychiatry);
- cognitive (mainly cognitive psychology and neuroscience).

In this paper it is not possible to discuss in detail all the research approaches mentioned above. For this reason, I present only some selected ideas, viewpoints and results of studies in various scientific fields. The presented conceptions are, in my opinion, crucial to the discussed subject and influence the studies from a neuroscientific approach.

1. Philosophical Approach

At the earlier stages of self-consciousness studies, only philosophers were engaged in explaining the problem of the essence of the self. The first records of self-consciousness and the first attempt to comprehend it go back to Indian philosopher Siddhartha Gautama. His meditations were centered on the control of consciousness. Afterwards, a number of philosophers considered the issue of what is the nature of the self. Later in the paper, some of their conceptions concerning the problem of self-consciousness are presented and it seems that the ideas of these philosophers have had a significant impact on researchers in other scientific disciplines and still play an essential role in recent debates.

Most of the work devoted to the issue of the self is connected to Rene Descartes. He is supposed to be one of the first thinkers whose concepts appeared to be particularly useful in contemporary studies on self-consciousness, although it seems that the origin of the well-known "cogito ergo sum" is related to the works of St. Augustine of Hippo. In his work *On the Trinity* in book X, St. Augustine considers the matter of how the mind knows itself. He came to the conclusion that every mind knew certainly three things concerning itself: firstly that it understands, secondly that it is, and thirdly that it lives. The starting point in understanding the existence of the self for Augustine was the question "who ever doubts that he himself lives, and remembers, and understands, and wills, and thinks, and knows, and judges?". Merely the existence of doubts can make us draw the conclusion that we exist, for the reason that consciousness (or self-consciousness) that can be doubting must exist.[4] According to St. Augustine, the self is a unity composed of memory, understanding and will.[5] Augustine claims that these three are one, meaning that they are one life, one mind, one essence. Following St. Augustine's way of thinking, Descartes bases his considerations on the statement that if I doubt, therefore I think, and, as a result, if I think, therefore I exist. In Meditation VI, Descartes stated a thesis called mind-body dualism: "I have a clear and distinct idea of myself as a thinking, non-extended thing, and a clear and distinct idea of body as an extended and non-thinking thing". Descartes distinguished between the mind,

[4] St. Augustine postulates: "Seeing that even if he doubts, he lives; if he doubts, he remembers why he doubts; if he doubts, he understands that he doubts; if he doubts, he wishes to be certain; if he doubts, he thinks; if he doubts, he knows that he does not know; if he doubts, he judges that he ought not to assent rashly. Whosoever therefore doubts about anything else, ought not to doubt of all these things; which if they were not, he would not be able to doubt of anything" in *On the Trinity*, book X, available from: <http://ebooks.adelaide.edu.au/>, accessed 10.06.12.

[5] As Augustine claims: „For I remember that I have memory and understanding, and will; and I understand that I understand, and will, and remember; and I will that I will, and remember, and understand; and I remember together my whole memory, and understanding, and will." in *On the Trinity*, book X, available from: <http://ebooks.adelaide.edu.au/>, accessed 10.06.12.

which did not possess various physical attributes, and the body. In addition he claimed that the mind was indivisible and the body could always be divided. Since the mind and body have different attributes, they must not be the same thing.[6]

From another perspective, the study on the self was conducted by John Locke. He postulates that for the subsistence of the person (which can be identified with the self) above all the existence of consciousness is the most essential.[7] He defines consciousness in "Essay Concerning Human Understanding" as "the perception of what passes in a man's own mind".[8] He argues that the self exists when we "can consider itself as itself, the same thinking thing, in different times and places".[9] For Locke, the crucial aspect of the self is the continuation of a person's memory content. He concludes that our knowledge of the self is intuitive.[10] He states that "when we see, hear, smell, taste, feel, meditate, or will anything, we know that we do so. Thus it is always as to our present sensations and perceptions: and by this everyone is to himself that which he calls self".[11] David

---

[6] I should note that as followers of Descartes, dualists are not a homogeneous group of thinkers. Philosophers have distinguished different varieties of dualism, for instance: substance dualists, interactionists, occasionalists, epiphenomenalists, property dualists. For example, Leibnitz postulates that the mind and the body exist in parallel and never interact with each other. While supporters of epiphenomenalism (e.g. T.H. Huxley, Jackson) state that mental states are caused by neural events, additionally mental states have no effects upon any neural events.

[7] „[...]only by that consciousness which is inseparable from thinking, and, as it seems to me, essential to it: it being impossible for any one to perceive without perceiving that he does perceive." [in:] J. Locke, *Essay Concerning Human Understanding*, Book II, chapter 1 (Of Ideas in general, and their Original), pas. 19, available from: <http://ebooks.adelaide.edu.au/>, accessed 10.06.12.

[8] *Ibid.*

[9] *Ibid.*

[10] Locke argues: "As for our own existence, we perceive it so plainly and so certainly, that it neither needs nor is capable of any proof. For nothing can be more evident to us than our own existence. I think, I reason, I feel pleasure and pain: can any of these be more evident to me than my own existence? If I doubt of all other things, that very doubt makes me perceive my own existence, and will not suffer me to doubt of that." [in:] J. Locke, *Essay Concerning Human Understanding,* Book II, chapter XXVII, pas. 9, available from: http://ebooks.adelaide.edu.au/, accessed 10.06.12.

[11] *Ibid.*

Hume represents nearly the same point of view. He argues that self--consciousness states could be understood as "a bundle or collection of different perceptions, which succeed each other with an inconceivable rapidity, and are in a perpetual flux and movement".[12] According to his standpoint, in every moment of our lives, we cannot find a second when we are without a perception. That is the reason, in his opinion, why we are unable to experience the self when we direct our attention inwards. Hume believes that while investigating the self we "can observe any thing but the perception".[13] So he assumes that the self-consciousness states have to be identified with different perceptions (heat, cold, light, pain), which we experience in present moment. Hume's conception is referred to as the bundle theory. Another traditional philosophical view which still dominates the recent debates is that of Immanuel Kant. He claims that the self is only a condition of the possibility of experience. Kant postulates that there has to be a self as a precondition of experience. If the self does not exist then we cannot explain the unity of our experiences.[14] Kant states that "it must be possible for the 'I think' to accompany all of my (*conscious*) representations, for otherwise [...] (*they*) would be nothing to me".[15]

The arguments derived from the works of the above-mentioned philosophers have frequently appeared in the recent debate over the phenomenon of self-consciousness. For instance, Daniel Dennett defines the self as a centre of narrative gravity. Dennett argues that a self is an abstract object, a theorist's fiction. He compares the self to the centre of gravity in physic "has no mass; it has no color; it has no physical properties at all, except for spatio-temporal location".[16] On

---

[12] D. Hume, *A Treatise of Human Nature*, Book I, IV *Of the Sceptical and Other Systems of Philosophy*, Sect. vi. (Of Personal Identity), available from: < http://ebooks.adelaide.edu.au/>, accessed 10.06.12.

[13] *Ibid.*

[14] I. Kant, *Critique of Pure Reason*, translated by J. M. D. Meiklejohn, Second Part, First Division (Transcendental Analytic), available at <http://ebooks.adelaide.edu.au/>, accessed 10.06.12.

[15] *Ibid.*

[16] D. Dennett, *Consciousness Explained*, Little, Brown, Boston 1991.

the other hand, for example, Thomas Nagel states that any person has an objective self as a non-reducible entity.[17]

2. Psychological Approach

The psychologist Wilhelm Wundt brought a new insight into the investigation of the problem of self-consciousness. He applied empirical methods for the first time to studies of consciousness and, as a result, moved consciousness out of the purely philosophical investigation sphere. Wundt proposed using introspection while investigating the self. Partly on the basis of Wundt's works, the psychologist William James built his theory which still plays significant role in scientific debate on self-consciousness. William James postulates that consciousness is just a product of brain activity and thus rejects dualism. In his opinion, the self consists of the self as knower or the 'I' and the self as known or the 'me'. The 'I' is conscious in itself (James calls it the pure Ego), while the 'me' is what the 'I' can be conscious. The 'me' is composed of the material, social and spiritual self. According to James, self-consciousness

> involves a stream of thought, each part of which as 'I' can 1) remember those which went before, and know the things they knew; and 2) emphasize and care paramountly for certain ones among them as 'me', and appropriate to these the rest. The nucleus of the 'me' is always the bodily existence felt to be present at the time.[...]This me is an empirical aggregate of things objectively known.[18]

---

[17] As noticed Albert Newen in *Self* [in:] *Encyclopedia of Neuroscience,* eds. M.C. Binder, N. Hirokawa, U. Windhorst, Springer-Verlag, Berlin–Heidelberg 2008, pp. 3635–3638.

[18] W. James, *The Consciousness of Self,* [in:] *The Principles of Psychology*, available from <http://ebooks.adelaide.edu.au/>, accessed 10.06.12.

James's distinction (between the 'I' and the 'me') is often adopted by other researchers. Sometimes it is even argued that, from the time when William James provided a distinction between different kinds of the self, philosophers, psychologists and neuroscientists try to adapt, extend, and redefine this concept.

Just like William James in psychology, Karl Jaspers in clinical disciplines, above all psychiatry, made the main contribution to understanding the essence of the self. In the framework of comparative study based on difference in experiencing of self-consciousness by healthy persons and schizophrenic patients (with the various self-disturbances), he suggested the following experiential modes:

- vitality – an awareness of being alive,
- activity – an awareness of one's own behaviours and abilities,
- unity – coherence of thinking and feeling,
- identity,
- demarcation – the ability to distinct yourself from the outside world.[19]

The self-disturbance disease may be caused by the disorder in one or more above-mentioned modes. The introduction of modes proposed by Jaspers is popular among scientists who deal with personality disorders.

3. Neurocognitive Approach

Contemporary theories of self-consciousness are focused on the knowledge gained from cognitive science. The investigation of the natural basis of self-consciousness with the methods of neuroscience gener-

[19] Cf. T. Kircher, A.S. David, *Self-Consciousness: An Integrative Approach From Philosophy, Psychopathology and the Neurosciences*, [in:] eds. T. Kircher, A.S. David, *The Self in Neuroscience and Psychiatry*, Cambridge University Press, Cambridge 2003, pp. 459–460.

ally goes in two main directions. The first is centred upon ontogenetic processes, including evolutionary processes. The second aims to discover the neural correlates of self-consciousness. In each of these directions we can distinguish roughly two approaches to the problem of nature of the self. The first approach is based on the assumption that the self can be considered as a group of various components. This group has to comprise a "minimal" sense of the self. Usually in this approach a list of features is presented which the person must have when she is self-conscious.[20] In regards to the complexity of the self, the studies are often concentrated on one of the aspects of self-consciousness (e.g. self-regulation or self-agency). The second approach is based on the postulation that it is possible to investigate the whole problem using a coherent concept of the self. Because of the wide range of empirical studies, it is impossible to present in this paper even a fraction of them and this is why only some of them are discussed here.

The studies from an ontogenetic perspective were started by executing the mirror test. Currently it is commonly believed that the person is self-conscious if she can pass the mirror test. The original study was conducted on a group of chimpanzees. The mirror was placed in the chimpanzees' cage and, when the animals got accustomed to their mirror images, the mirror was taken away. Afterwards the chimpanzees were anaesthetized and red marks were put on their muzzles (in a place invisible for them, above the eyebrow). After some time, the mirror was again placed in the animal's cage. It turned out that chimpanzees touched the marks on their muzzles.[21] The mirror test carried

---

[20] An example could be a list proposed by Albert Newen: perspectivity, agency, and mineness. Newen states: "Perspectivity includes the first-person-perspective we have in human perception […]. Agency means the feeling that we are causing an action, the feeling that the action is performed because we want to perform it. Mineness can be characterized as the feeling that a bodily part (an arm, a leg) belongs to me, but also the feeling that a thought is my thought." in *Self* [in:] *Encyclopedia of Neuroscience,* eds. M.C. Binder, N. Hirokawa, U. Windhorst, Springer-Verlag, Berlin–Heidelberg 2008, pp. 3635–3638. Another example is the classification proposed by Lukas Heydrich: self-location, first-person perspective and self-identification.

[21] G. Gallup, *Chimpanzees: Self Recognition,* "Science" 1970, vol. 167, no. 3914, pp. 86–87.

out on human beings reveals that infants can pass the test at the age of eighteen months and this is in spite of the fact that children recognize relatives and strangers in early infancy. There exists, however, an inconsistency in the child's perception of the surroundings. On the one hand, an infant can identify third persons and distinguish the movements of its body in the mirror but, on the other, it cannot pass the mirror test. This contradiction can be explained by the assumption that some neurocognitive components of self-consciousness develop at a different age.[22] Evidently the mirror test is connected to the ability of self-recognition. It is supposed that the part of the brain responsible for self-recognition is the ventromedial regions of the right prefrontal cortex. The proposed localization results from the fact that this part of the brain is more developed in humans compared to primates. In addition, it is formed in children as one of the later parts and empirical studies confirm this supposition (for instance the studies on patients suffered from asomatognosia or the studies in which the anesthesia of the right hemisphere is used).[23]

The second direction of study concerning the creation of models of the self demands the establishment of what the correlation is between consciousness and self-consciousness. Is self-consciousness one of the aspects of consciousness? Is self-consciousness essential to consciousness? These matters are crucial for all models of consciousness. Each of them is partially related to the problem of the self and here only some of the existing models are discussed. The most radical ideas are presented by Paul Churchland. He believes that exploring

---

[22] M. Lewis, M. Sullivan, et al., *Self-development and Self-conscious Emotions*, "Child Development" 1989, vol. 60, pp. 146–56.

[23] Asomatognosia is a disorder of body awareness caused by neurological damage (S. Dieguez., F. Staub, J. Bogousslavsky, *Asomatognosia,* [in:] *The Behavioral and Cognitive Neurology of Stroke,* eds. O. Goderfroy, J. Bogousslavsky, Cambridge University Press, Cambridge 2007, pp. 215–253.). Cf. e.g. T.E. Feinberg, R. Shapiro, *Misidentification-reduplication and the Right Hemisphere*, "Neuropsychiatry, Neuropsychology, and Behavioral Neurology" 1989, vol. 2, pp. 39–48; T.E. Feinberg, L.D. Haber, N.E. Leeds, *Verbal Asomatognosia,* "Neurology" 1990, vol. 40, pp. 1391–4.; K.J. Meador, D.W. Loring, et al., *Anosognosia and Asomatognosia During Intracarotid Amobarbital inactivation,* "Neurology" 2000, vol. 55, pp. 816–20.

the nature of consciousness (as well as self-consciousness) is actually pointless since the conception of conscious states will be eliminated in the future. When we become acquainted with all the functions of our brain and their mechanisms, we may come to the conclusion that consciousness is only a subjective illusion because, objectively, only the impulses conducted by neuronal cells exist.[24] The argumentation of Antonio Damasio takes another direction as he claims that there is a strong correlation between specific mental states and the activity of certain area of the brain (group of neurons or even single cells). Damasio states that in order to understand consciousness, one has to resolve two problems, so called: the "movie in the brain" experience[25] and the self.[26] Damasio postulates that brain cells are "intentional". The brain has an inner representation of the body itself, as well as the objects interacting with the body in external world. He also believes that the key to the self is the representation of the continuity of the organism. The postulations stated by Damasio resulted in the creation of the "brain map" concept. Such a map comprises the information about what area of the brain is responsible for definite cognitive ability, in other words it defines the neuronal basis of the "movie in the brain". Also in his studies he argues that an understanding of all of the functions of the brain leads us to the revelation of self essence.[27] It is also important to mention at this point the contribution of Bernard Baars. He proposes the 'global workspace' theory, which explains inter alia the origin of conscious processes. This theory is based on the assumption that the central neural system is responsible for the information processing and its transfer to the conscious plane. The different types of information, external stimuli, and unconscious processes compete with each other for access to a global workspace. The main

---

[24] P.M. Churchland, *Eliminative Materialism and the Propositional Attitudes*, "Journal of Philosophy" 1981, vol. 78, pp. 67–90.

[25] He means "how a number of sensory inputs are transformed into the continuous flow of sensations of the mind".

[26] He claims "how the sense of 'owning' that movie comes to be".

[27] A. Damasio, *The Feeling of What Happens,* Harcourt Brace, New York 1999.

functions of the global workspace are coordination, control and information processing. Baars suggests that the global workspace is located in the extended-reticular-thalamic-activating system (ERTAS), which is composed of the reticular formation, the thalamus, and the "diffuse thalamic projection system".[28] He also postulates that consciousness is generally accompanied by a subjective sense of self.[29] In agreement with earlier considerations by Baars, John Searle claims that consciousness consists of inner, qualitative, subjective states and processes. He believes that consciousness is composed of higher-level processes realized in the structure of the brain. According to Searle "conscious states, therefore, have what we might call a 'first-person ontology'. That is, they exist only from the point of view of some agent or organism or animal or self that has them".[30] He summarizes, that because there exists a first-person ontology, consciousness cannot be reduced to third-person phenomena in the way typical for other natural phenomena (for instance heat or liquidity). Nevertheless, Searle believes that consciousness, as well as conscious processes, can be treated on the same footing as biological processes.[31]

A completely different way of thinking about the discussed subject is presented by Jonathan C. W. Edwards. He claims that the self may reside in every neuron. He postulates that this hypothesis solves the problem of how a single conscious experience arises from an array of diverse neurons, the so-called binding problem.[32] Edwards notes that each neuron can receive a huge quantity of information every second as every neuron has thousands of synapses. In Edwards' opinion, this feature may be fundamental for understanding the complexity of

---

[28] B.J. Baars, *A Cognitive Theory of Consciousness,* Cambridge University Press, New York 1988.

[29] Baars argues that we can understanding subjectivity using global workspace theory and concept of the resurrection of the observing self.

[30] J. Searle, *Mind, Language and Society: Philosophy in the Real World*, Weidenfeld & Nicolson, London 1999.

[31] Cf. *idem*, *The Rediscovery of the Mind*, MIT Press, Cambridge, Mass. 1992.

[32] J. Edwards, *Is Consciousness Only a Property of Individual Cells*, "Journal of Consciousness Studies" 2005, vol. 4–5, pp. 60–76.

conscious experience. He considers that "a limited form of consciousness may reside in other cells of the body. [...] while each neuron may be aware of a wide range of stimuli, it is likely that 99% of volitional acts by the 'Person' are 'decided upon' elsewhere in the brain".[33] That is the reason why most of our decisions and thoughts are unconscious. Nevertheless, this does not influence the matter that "each neuron has access to the 'story of the self' concocted by neurons in the left hemisphere and falls under the delusion that it is this self".[34]

In recent debates, an essential standpoint has also been presented by Thomas Metzinger. In his theory of self-consciousness, the human being is treated like a neural machine and has the unique ability to create a self-model. He believes that "there are not such things as selves".[35] He argues that there exists the phenomenal self, which is not a thing but an ongoing process. In Metzinger's opinion the self is the content of a "transparent self-model".[36]

Quite on the contrary, Galen Strawson concludes that there are many short-lived and successive selves, in the case of individual humans. He proposed that "the self tends to be figured as:

> [1] a subject of experience, a conscious feeler and thinker;
> [2] a thing, in some interestingly robust sense;
> [3] a mental thing, in some sense;
> [4] a thing that is single at any given time, and during any unified or hiatus-free period of experience;
> [5] a persisting thing, a thing that continues to exist across hiatuses in experience;
> [6] an agent;
> [7] as something that has a certain character or personality".[37]

---

[33] *Ibid.*
[34] *Ibid.*
[35] T. Metzinger, *Being No One*, MIT Press, Cambridge, Mass. 2003.
[36] *Ibid.*
[37] G. Strawson, *The Self and the SESMET*, "Journal of Consciousness Studies" 1999, vol. 4, pp. 99–135.

Strawson proposed the above list in order to give the basic conceptual structure of our sense of the self. The model proposed by Arnold Trehub partly supports Strawson's ideas about the nature of the self and Trehub suggests that the existence of a post-retinal dynamic buffer in the brain, which he named the retinoid system, is substantial for the self.[38] Trehub argues that the autaptic neurons[39] at the self locus of the retinoid structure constitute the core biophysical substrate for our sense of self. He postulates that the retinoid system embodies conscious experience of egocentric space and the self is its neuronal origin. In his opinion "in order to have a self, we must have an I-token with intact synaptic coupling to the self-locus neurons in the retinoid system".[40]

One of the recent models of the self has been proposed by Stuart Hameroff. He presents a new model of the neural correlates of consciousness "as a mobile integrator moving through dendritic/somatic input/integration layers of the brain's neurocomputational networks".[41] Here Hameroff refers to David Chalmers' concepts. Chalmers argues that there are two types of the problems regarding consciousness: "easy" and "hard". The "easy" problems concern the unconscious activity of the brain in response to different external stimuli (e.g. control behaviour). These non-conscious cognitive functions are often called "auto-pilot" modes, because they are mostly habitual while the "hard" problems relate to the question of how the brain produces conscious experience. However, in Hameroff's opinion, a distinction between non-conscious auto-pilot modes and conscious experience is not essential. He argues that a distinction be-

[38] Trehub states that "its putative structural and dynamic properties enable it to register and appropriately integrate disparate foveal stimuli into an egocentric representation of an extended 3D frontal scene, as well as perform many other useful perceptual and higher cognitive functions" in A. Trehub, *Space, Self, and the Theater of Consciousness*, "Consciousness and Cognition" 2007, vol. 16 (2), pp. 310–330.

[39] Self-synapsing neurons.

[40] A. Trehub, *Space, Self, and the Theater of Consciousness*, "Consciousness and Cognition" 2007, vol. 16 (2), pp. 310–330.

[41] S. Hameroff, *The "Conscious Pilot"—Dendritic Synchrony Moves Through the Brain to Mediate Consciousness,* "Journal of Biological Physics" 2009, vol. 36 (1), pp. 71–93.

tween "cognitive neurocomputation which at any given moment is, or is not, accompanied by some added fleeting feature (the 'conscious pilot') which conveys conscious experience and choice" is more useful.[42] The conscious pilot in Hameroff's model refers to spatiotemporal envelopes of dendritic gamma synchrony (structurally defined by gap-junction-connected dendritic webs). These waves propagate through the brain as a conscious agent which can experience and control otherwise non-conscious (auto-pilot) cognitive functions. He considers that the auto-pilot is neurocomputational and is mediated by axonal firings and synaptic transmissions in feedforward and feedback neuronal networks. Although the conscious pilot is "synchronized collective integration (dendritic webs) moving and evolving sideways through input/integration layers of the same neuronal network architecture".[43]

Conclusion

In summarizing the contribution of neuroscience to the solution of the problem of the self, we can unquestionably attain the following conclusions: the phenomenon of self-consciousness does not arise from a single neural process and the self is not a monolithic entity. Also, from other approaches, it is obvious that we are still far from a complete understanding of the essence of self-consciousness. It is known that there is a significant difference between our knowledge about the brain, people's behaviour, and our knowledge about cognitive abilities and their origin.

In this context, the key question is how we can investigate the natural foundations of the self with the methods of cognitive neuroscience. It seems that the most appropriate way to study the nature of self-consciousness from the perspective of neuroscience is to conduct

---

[42] *Ibid.*

[43] *Ibid.*

research in the following way: firstly, on the basis of the philosophical assumptions, the concept of the self is created; secondly, the "components" of the self are listed; thirdly, each of the "components" of the self is investigated using empirical methods, and the last stage is the verification of the concept. Each of these stages has its difficulties to overcome and here I discuss only some of them. First of all it is worth examining what philosophical foundations a model of the self must have. It appears that it is helpful to survey the philosophical assumptions on which the earlier models are founded. While analyzing previous neuroscience concepts of self-consciousness, it seems that most of them are based on assumptions which are not perfectly appropriate either to the problem of the self, or to the methods of cognitive neuroscience. The recent debate on the mind-body problem may serve as evidence. From the self-consciousness perspective, the mind-body problem is quite crucial. In neuroscience studies, one of the solutions to this problem (dualism or monism) is assumed and, from a neuroscience perspective, neither of them is without reservation. The main objection to dualism is that the existence of non-physical mind implies that it is impossible to find a naturalistic explanation of mental facts. It seems that in case of neuroscience the acceptance of such a research standpoint is not appropriate since, instead of extending the realm of naturalistic explanations, it limits explanatory scope by assuming that there are in principle facts beyond explanation. On the other hand, the interpretation proposed by monists is also rather unsatisfactory.[44] It appears that each of the monistic theories have some gaps and inaccuracies in explanations of the existence and the origin of mental events. As a consequence it appears that it is well justified to continue to explore new philosophical approaches to the questions concerning self-consciousness. Another difficulty is how to construct the list of self-consciousness "components". Certainly, the earliest re-

[44] It is worth mentioning that there are a wide range of monistic interpretations of the mind and body problem. For instance, the followers of eliminative materialism (eliminativism) postulate that mental states do not exist. According to the identity theorists, mental states have to be treated only with correlation to particular physical states.

search in different scientific fields (such as psychology, psychiatry) seems to be very supportive. These criteria, on the one hand, have to be coherent and, on the other hand, it must be possible to verify them by empirical studies. Finally it is worth mentioning that in the case of research on the self, the methodology of studies has a fundamental importance and, in particular, the perspective from which the studies are conducted is crucial. In the philosophical area, the so-called first--person perspective is used and the experience of cognitive states is evaluated by the person they relate to. In the event of the acceptance of the first-person perspective, two central problems emerge. The first is how to determine the correlation between the thinking subject and the external surrounding environment, and its influence. The second problem is connected to one of the aspects of self-consciousness – that of self-knowledge. The question is in what way the person has access to her self-knowledge. Neuroscience research adopted quite a different perspective which is based on the assumption that only states that exist objectively and interpedently of the thinking subject must be investigated and this is the so-called the third-person perspective. In this perspective, the mental states of the subject are not evaluated, only the external manifestations of these states are taken into consideration (e.g. behaviour).[45] The fundamental problem related to the third-person perspective is that mental features are elusive and none of their characteristics are noticeable. Between the above-mentioned perspectives, the second-person perspective resides. Such a perspective is used in psychology and psychiatry. Each of these research perspectives has their flaws. It seems that in studies on the nature of the self it is not necessary to confine ourselves to only one of them. In the case of neuroscience, in my opinion, the acceptance of only the third--person perspective is not appropriate. The reason is that the experience of self-consciousness is, by its nature, subjective. That is why it cannot be examined simply by objective research methods. In order

[45] This perspective is used in research conducting by e.g. Dennett, Damasio, Churchland.

to comprehend the essence of the self, the first-person and third-person perspective should be taken into consideration while conducting the research.

In conclusion, it is worth mentioning that, in some cases, even perfectly constructed research will be not able to bring us closer to the full cognition of the nature of the self. Because of the fact that while investigating the neural correlates of self-consciousness (the dependence between reactions in the brain and mental states), in practice we may find only the correlations between our knowledge about reactions in the brain and our knowledge about mental events. To sum up, the current state of self-consciousness research raises significantly more questions than it delivers answers. All of this suggests that the hardest challenges in this field are still to be faced.

Mateusz Hohol, Piotr Urbańczyk
The Pontifical University of John Paul II
Copernicus Center for Interdisciplinary Studies

Self-Deception
Between Philosophy and Cognitive Neuroscience[1]

Introduction

There is no doubt that the question of self-deception constitutes an interdisciplinary problem. Even within the confines of philosophy it is an object of research in various fields such as epistemology, philosophical anthropology, philosophy of mind, ethics and logic. Psychological research on self-deception has been conducted for a number of decades and, most recently, it has attracted the attention of biologists. This paper aims to discuss three models of self-deception which have been created within the framework of cognitive neuroscience. We are, however, aware of the philosophical origins of the problem, which are presented in the first part of this study. We believe that self-deception is a typical problem of philosophy *in* neuroscience. The author of the idea of philosophy *in* science – Michael Heller – delineates its three fields as:

> (A) the influence of philosophical ideas on the development and evolution of scientific theories; (B) traditional philosophical problems

[1] This paper was written within the research grant "The Limits of Scientific Explanation" sponsored by the John Templeton Foundation.

> intertwined with empirical theories; (C) philosophical reflection over some assumptions of the empirical sciences.[2]

Nevertheless, he adds that nothing stands in the way of extending this set of fields. For example, Łukasz Kurek mentions the area specific for neurophilosophy, which is (D) "implementation of the achievements of neuroscience in reflections concerning traditional philosophical problems".[3] We will try to show that the problem of self-deception applies to all of these areas.

1. Self-Deception in Philosophical Debates

Self-deception has enjoyed the unflagging interest of philosophers long before it attracted the attention of neuroscientists. The philosophical wrangling over the topic has generated a large and growing body of literature of its own. As usually happens in such discussions, there is no general agreement as to the nature, definition or even the most paradigmatic cases of the phenomenon in question. Virtually all attempts to characterize both the state of being self-deceived and the process of self-deception have been criticized for lacking in accuracy and failing to capture all aspects of the everyday usage of the term. It is not our aim to describe the whole philosophical debate concerning self-deception but rather to draw on its main points.

---

[2] M. Heller, *How is Philosophy in Science Possible?*, translated by B. Brożek, [in:] *Philosophy in Science. Methods and Applications*, eds. B. Brożek, J. Mączka, W.P. Grygiel, Copernicus Center Press, Kraków 2011, p. 15.

[3] Ł. Kurek, *Neurofilozofia jako filozofia w kontekście nauki*, [in:] *Oblicza racjonalności. Wokół myśli Michała Hellera*, eds. B. Brożek, J. Mączka, W.P. Grygiel, M.L. Hohol, Copernicus Center Press, Kraków 2011, pp. 83–82.

1.1. The Standard Account and Its Two Paradoxes

The most common approach to the problem is based on the assumption that self-deception is somehow isomorphic with stereotypical interpersonal deception. The latter occurs when one persuades someone else to believe that some proposition *p* is true, while knowing, or at least truly believing, that *p* is false. To model self-deception we assume that the deceiver and the deceived are the same person; i.e., self--deceivers know or believe that *p* is false and intentionally get themselves to believe that *p* is true.

The above model immediately leads to two paradoxes. The first of them is concerned with the state of self-deception. The model requires that the self-deceiver holds two contradictory beliefs (*p* and not-*p*) at the same time, which is thought to be an impossible state of mind. The second paradox focuses on the process (or activity) of self--deception. For self-deceivers to succeed, they have to be unaware of their own deceitful strategy. Going back to the interpersonal deception, how could I be deceived by someone, if I know that he wants to deceive me? Self-deceivers persuade themselves to believe that *p* is true while believing the opposite seems to be involved in some kind of self-defeating project. Mele called these paradoxes *static* and *dynamic* (or *strategic*) puzzles respectively.[4]

These "puzzles" have produced a whole gamut of various responses. Those sceptical about self-deception (the minority of philosophers) think that the very notion of self-deception is simply implausible, Mary Haight being an excellent example. She argues that since the possibility of self-deception leads to contradiction, it should be rejected. The paradoxical nature of the conception "is real, so the name – whatever we mean by it – cannot literally mean that".[5] Nevertheless,

---

[4] See A. Mele, *Irrationality: An Essay on Akrasia, Self-Deception, Self-Control*, Oxford University Press, Oxford 1987, chapters 9–10; *idem*, *Real Self-Deception,* "Behavioral and Brain Sciences", 1997, vol. 20, p. 92 and *idem*, *Self-Deception Unmasked*, Princeton University Press, Princeton 2001, pp. 7–8.

[5] M.R. Haight, *A Study of Self-Deception*, The Harvester Press, Brighton 1980, p. 1.

most philosophers have claimed that self-deception is possible and hence have tried to find a way to subvert or avoid the paradoxes mentioned above. Roughly speaking, two ways out have been proposed. Firstly, one can justify how it is possible for the self-deceiver to hold contradictory beliefs (and this is mostly done by denying that they are held simultaneously, consciously or in the same "mental partition"). The main assumption that is made here is that self-deception is intentional, thereat such approaches were dubbed intentionalist. Secondly, one can simply reject the model based on interpersonal deception and this is precisely what anti-intentionalists do.

1.2 Split Selves

Intentionalists generally agree that self-deception can be modelled on interpersonal deception and they admit that the phenomenon involves some sort of intentional acts. Nevertheless, there is no general agreement as to the requirement for the holding of contradictory beliefs.[6] Despite this disagreement, they utilize some common means of avoiding the paradoxes that self-deception produces. These means consist of some kind of partitioning of consciousness. To separate two inconsistent beliefs, or at least to insulate self-deceivers from their intention to acquire a belief that is known to be false, this mental partitioning may take the form of a temporal or psychological division or even a combination of them.

According to the first account, self-deception is a long-term and gradual event. No act of self-deception achieves its aim instantaneously and there must be sufficient time between the outset and the completion of the process of the deception to allow for the belief to be acquired.[7] An intention that causes belief acquisition is fully con-

[6] Cf. I. Deweese-Boyd, *Self-Deception*, [in:] *The Stanford Encyclopedia of Philosophy*, ed. E.N. Zalta, Stanford 2012, <http://plato.stanford.edu/archives/spr2012/entries/self-deception/>.

[7] R.A. Sorensen, *Self-Deception and Scattered Events*, "Mind" 1985, vol. 373, p. 69.

scious at the time that it is formulated. Nonetheless, while self-deceivers proceed in implementing it, they can forget their original motivation. It is argued that one can be unconscious of the intentions that gave rise to her actions, although she was conscious of them at the beginning.[8] Hence, in such a view, there is no question of contradictory beliefs being held simultaneously nor of an awareness and ignorance of deceitful strategy simultaneously occurring in the mind. Self-deceivers can believe that *p* at the beginning of the process and intentionally get themselves to believe the opposite, with the result that they believe that not-*p* at its completion.

There is no doubt that the time-lag account succeeds in avoiding paradoxes of self-deception. There are several philosophers, however, who claim that the idea of one deceiving her "future self" generate its own puzzles anyway[9] and – foremost – it is not capable of capturing all cases of self-deception and leaves most of them untreated.

Some intentionalists have taken the analogy with interpersonal deception quite seriously and have attempted to allow the possibility of self-deception by dividing up the mind. The theory with perhaps the strongest literal connotations of the divided self was proposed by Amélie Rorty. She explains this alleged inner division of self-deceivers in terms of the autonomous subsystems of habits of rationality, inference, calculation, imagination, expectation, etc. In her pluralistic conception of the self, one person constitutes in fact a multitude of such subagencies. Agents vary in the ways their subagencies are structured – "some form quite strictly unified systems with strong centralization, others are quite loosely connected, with different habits coming to the fore in response to situational variations".[10] Surprisingly enough, in

[8] Cf. J.L. Bermúdez, *Self-Deception, Intentions and Contradictory Beliefs*, "Analysis" 2000, vol. 60(4), p. 314.

[9] See B.P. McLaughlin, *Exploring the Possibility of Self-Deception*, [in:] *Perspectives on Self-Deception*, eds. B.P. McLaughlin, A.O. Rorty, University of California Press, Berkeley 1988, pp. 36–38; M. Johnston, *Self-Deception and the Nature of Mind*, [in:] *Ibid.*, pp. 76–78.

[10] A.O. Rorty, *Self-Deception, Akrasia and Irrationality*, [in:] *The Multiple Self*, ed. J. Elster, Cambridge University Press, Cambridge 1985, p. 130.

her view self-deception concerns the former case, i.e., it concerns persons whose subsystems are centralized. The phenomenon in question occurs when one of the subsystems takes a dominant and integrating function over the others.[11]

A similar account was offered by David Pears. He argues for distinguishing two parts in self-deceivers – the subject and the object of deception. He also claims that the distinction between these parts cannot be based on the line dividing what is conscious and unconscious and the insulation is rather functional. Both of the parts could be rational as well. Nonetheless, the subject of deception is a limited subsystem of the deceived part, which plays the role of a centre of agency within the whole person.[12]

According to the weaker account proposed by Donald Davidson, self-deception is possible thanks to the division of the mind into sets of states and processes. These sets are semi-independent (in a sense that usual logical and epistemic relations between them are broken down), although it is not claimed that they form separate entities that have distinct functions in mind. Davidson even claims that phrases like "mind partitioning" could be "misleading if they suggest that what belongs to one division of the mind cannot belong to another"[13] and the boundaries between parts of the mind are just "conceptual aids to the coherent description of genuine irrationalities".[14] Hence, in this view, the metaphor of the multiple self could be avoided in a way.

Another intentionalist approach was put forward by Demos. In his explanation of the phenomenon, he utilizes a notion of two different level of awareness – simple awareness and the awareness together with attending or noticing. Self-deception in acquiring a belief that *p* is possible, because self-deceiver although aware of her belief that

[11] See *ibid.*, p. 131.

[12] See D. Pears, *The Goals and Strategies of Self-Deception*, [in:] *Ibid.*, pp. 69–77.

[13] D. Davidson, *Paradoxes of Irrationality*, [in] *Philosophical Essays on Freud*, eds. R. Wollheim, J. Hopkins, Cambridge University Press, Cambridge 1982, reprinted in: D. Davidson, *Problems of Rationality*, Oxford University Press, Oxford, 2004, p. 181.

[14] D. Davidson, *Deception and Division*, [in:] *The Multiple Self*, *op. cit.*, p. 91.

not-*p* does not attend to it nor focus her attention on it. Demos compares this to situation, when one does not notice her headache while her attention is absorbed by a movie, for instance.[15]

Demos' account has been taken in the first serious experimental demonstrations on self-deception.[16] Two psychologists – Gur and Sackeim – used his idea that in the self-deceived individual two contradictory beliefs are held on different levels of awareness. They conceptualized this by listing four conditions that are necessary and sufficient for ascribing self-deception to a phenomenon:

1. The individual holds two contradictory beliefs (that *p* and not-*p*).
2. These two contradictory beliefs are held simultaneously.
3. The individual is not aware of holding one of the beliefs.
4. The act that determines which belief is and which belief is not subject to awareness is a motivated act.[17]

To test these ideas, Gur and Sackeim adopted a form of voice-recognition experiment. Participants were asked to discriminate between taped samples of others' voices and their own. Those who previously had failed in the cognitive test were slowest in recognize their own voices and denied hearing their voices more often than subjects who had succeed. Nevertheless, records of their physiological responses (galvanic skin responses) suggested that recognition did occur. Gur and Sackeim argued that this misinterpretation fulfills the criteria for ascribing self-deception,[18] although their experiment has raised several objections. First and foremost, it seems controversial to state that

---

[15] See R. Demos, *Lying to Oneself*, "The Journal of Philosophy" 1960, vol. 18, p. 593.
[16] Cf. T. Sturm, *The Self Between Philosophy and Psychology: The Case of Self-Deception*, [in] *Psychology's Territories. Historical and Contemporary Perspectives from Different Disciplines*, eds. M.G. Ash, T. Sturm, Lawrence Erlbaum Associates, Mahwah-London 2007, p. 181.
[17] R.C. Gur, H.A. Sackeim, *Self-Deception: A Concept in Search of a Phenomenon*, "Journal of Personality and Social Psychology" 1979, vol. 37, p. 149.
[18] Cf. *ibid.*

a skin response could serve as an indicator of holding of a belief[19] (in this case – contradictory to a belief reported). Despite this objection, experiments on self-deception with unconscious beliefs operationalized in the same manner have been continued.[20]

1.3 Motivational Bias

Several philosophers have chosen to argue that self-deception is unintentional and thereby they have broken away from an important aspect of the model of self-deception based on its interpersonal analogue, with perhaps the most representative example being Mele. Strictly speaking, he argues that self-deceivers can rarely act with the intention of deceiving themselves, but the vast majority of cases of self-deception are cases of unintentional deception.[21] According to his approach, the phenomenon should be explained without the assistant of "mental exotica" of subagencies, mental partitioning, unconscious beliefs, etc. First of all, the requirement that a self-deceiver believes that *p* while knowing that not-*p* has been dropped out on his view, so there is no need to state that a person holds two contradictory beliefs. As a result of a self-deceptive act, one is just mistaken in believing or believes falsely. In addition, he claims that not all cases of self-deception involve the acquisition of a new belief. One might be self-deceived in maintaining a belief that she was not self-deceived in acquiring.

Mele proposes a hypothesis that desiring something to be true often exerts a biasing influence on what one believes and, in the case of self-deception, we deal with a sort of motivationally biased belief. The motivational bias leads to a manipulation of the data relevant to

---

[19] Cf. A. Mele, *Real Self-Deception*, *op. cit.*, p. 97. See also T. Sturm, *op. cit.*, pp. 181–182.

[20] E.g. J.E. Starek, C.F. Keating, *Self-Deception and Its Relationship to Success in Competition*, "Basic and Applied Social Psychology" 1991, vol. 12(2), pp. 145–155.

[21] See A. Mele, *Irrationality*, *op. cit.*, pp. 121–124.

the truth value of the objects of those beliefs. This manipulation could take the form of misinterpretation, attention shifting or selective evidence-gathering.

For the purposes of our further investigations it should be pointed out that some intentionalists utilize the concept of motivational bias as well. For example, Davidson admits that self-deceivers may be motivated by a desire to believe what they wish were the case and this inclination causes them to disregard the evidence for its falsity.[22] Nevertheless, he claims that self-deceivers must be aware of the evidence against motivationally biased belief and Mele cannot agree to this.

To summarize the anti-intentionalist account, let us note four conditions that – according to Mele – are sufficient for entering self-deception in acquiring a belief that *p*:

1. The belief that *p* which *S* [i.e., subject – M.H., P.U.] acquires is false.
2. *S* treats data relevant, or at least seemingly relevant, to the truth value of *p* in a motivationally biased way.
3. This biased treatment is a nondeviant cause of S's acquiring the belief that *p*.
4. The body of data possessed by *S* at the time provides greater warrant for not-*p* than for *p*.[23]

To justify these ideas he tries to deal with psychological literature.[24] His attempt, however, should be meant rather as a proposal of the conceptualization of the phenomenon, not as an empirical hypothesis.

[22] D. Davidson, *Deception and Division*, *op. cit.*, p. 87.
[23] See A. Mele, *Self-Deception Unmasked*, *op. cit.*, pp. 50–56; Cf. *idem*, *Irrationality*, *op. cit.*, 127–131.
[24] Cf. *idem*, *Real Self-Deception*, *op. cit.*

2. The Structure of Cognitive Neuroscience

As we have tried to show in the first part of this paper, studies on self-deception originally took the form of conceptual analyses mostly conducted by philosophers. In subsequent parts we will show how this issue is taken in the theories which can be categorized as fields of cognitive neuroscience. First, however, we will analyse what we understand by the theory of cognitive neuroscience.

The notion of the cognitive neuroscience is ambiguous. Moreover, it is difficult to determine what the structure of neurocognitive theory amounts to. The words of Valerie Gray Hardcastle can serve as a good illustration of this: "Brains are complicated and messy affairs; theories about brains share these same traits".[25] Furthermore, cognitive neuroscience is one of the youngest research fields that emerged from neurobiology in the late '90s. It seems to us, however, that it is possible to point out a couple of features which are unique to cognitive neuroscience.

Cognitive neuroscience is based on applying research methods which are typical for 'basic' neuroscience. Those methods are: single-cell recording, neuroimaging (fMRI, MEG, PET) and lesion-studies. The application of the methods mentioned above is extended with various types of behavioural experiments. The specificity of cognitive neuroscience is based on the fact that those methods are used not only for examining single neurons and more complicated neuronal structures, but also for examining the cognitive mechanisms of humans.[26] Cognitive neuroscientists often refer to comparative research concerning human brains and those of other primates. Nonetheless, this gamut of

[25] V. Gray Hardcastle, *Neurobiology*, [in:] *The Cambridge Companion to Philosophy of Biology*, eds. D. Hull, M. Ruse, Cambridge University Press, Cambridge 2008, p. 275.

[26] See W. Bechtel, *Epistemology of Evidence in Cognitive Neuroscience*, [in:] *Philosophy and the Life Sciences: A Reader*, eds. R. Skipper Jr., C. Allen, R.A. Ankeny, C.F. Craver, L. Darden, G. Mikkelson, R. Richardson, MIT Press, Mass. [in press], available online: <http://mechanism.ucsd.edu/epist.evidence.bechtel.july2004.pdf>.

methods is supplemented with several assumptions which have arisen from the adoption of the interpretative paradigm. We will return to this aspect later.

Another crucial feature of cognitive neuroscience is the fact that its theories operate at many levels of complexity. Depending on needs, theories could concern different levels, but in the most basic approach one can distinguish the following three: (1) level of single cells, (2) level of complex neuronal structures and (3) level of psychological phenomena. Roughly speaking, the last level concerns what may be called *the Mental*. It is easy to deduce that this level arouses much controversy and any understanding of it is largely determined by the adopted interpretational paradigm. Antti Revonsuo characterizes the third level in the following way:

> Cognitive neuroscience sees psychological levels (conceptualized as, e.g. "cognition," "information processing," "representation," "computation") as the higher levels of description, to be explained by referring to the neural and neurocomputational mechanisms residing at the lower levels. In this view, psychological phenomena are not explanatory autonomous, but neither are they eliminable – just like cytology is neither eliminable nor autonomous in the relation to biochemistry and molecular biology. Psychological properties are regarded as residing at a level of organization higher then neural properties, but nevertheless as being *micro-based properties* essentially in the same sense as other special-science properties.[27]

Even such a superficial description of cognitive neuroscience reveals its complexity and richness on the one hand and its problematical status from the viewpoint of methodology and philosophy of science on the other. Indeed, this discipline provokes one to reconsider well

---

[27] A. Revonsuo, *On the Nature of Explanation in Neurosciences*, [in:] *Theory and Method in the Neurosciences*, eds. P. Machamer, R. Grush, P. McLaughlin, University of Pittsburgh Press, Pittsburgh 2001, pp. 56.

known questions concerning, for example, scientific explanation[28] or the criteria of justification.[29]

Furthermore, the picture of cognitive neuroscience is complicated by the fact that its representatives adapt different interpretational paradigms. By interpretational paradigm we mean the attitude, or rather the meta-theory, which consists of: rules governing the construction of experiments, methods of interpretation of experimental data, the basic objective of the research, methods of generation of the scientific explanations, methods of understanding of basic concepts of, e.g. the mind, adopted anthropological and philosophical assumptions, etc.[30] In our opinion, such paradigms are, among others, computationalism (which is still present), evolutional psychology as well as the 'embodied-embedded mind' paradigm.

The supporters of computationalism used to interpret experimental data in terms of information processing. The level 3 is treated by them as strictly algorithmic, but implemented in the biological hardware (in the distinction mentioned above the latter corresponds to levels (1) and (2)).[31] The main objective of the computationalists is, above all, *explanation* through the discovery of computational mechanisms and creating of cognitive architectures.

Evolutionary psychologists adopt the basic postulate of the computationalists concerning the level (3), namely, the computability of the mind. This demand usually takes form of the strong modular theory of mind, known as Massive Mental Modularity.[32] Evolutionary psycholo-

---

[28] Cf. M. Miłkowski, *Theoretical Unification and the Neural Engineering Framework*, this volume; C.F. Craver, *Explaining the Brain. Mechanisms and the Mosaic Unity of Neuroscience*, Oxford University Press, Oxford-New York 2007.

[29] Cf. B. Brożek, *Philosophy in Neuroscience*, [in:] *Philosophy in Science. Methods and Aplications*, eds. B. Brożek, J. Mączka, W.P. Grygiel, Copernicus Center Press, Kraków 2011, pp. 163–188.

[30] Cf. *ibid.*, pp. 181–183.

[31] Cf. J.R. Anderson, *Methodologies for Studying Human Knowledge*, "Behavioral and Brain Sciences" 1987, vol. 10, pp. 467–505.

[32] The forerunner of this approach is Jerry Fodor; cf. *idem*, *The Modularity of Mind*, MIT Press, Mass.-London 1983.

gists also emphasize the evolutionary origin of level (3).[33] That is why they – in their pattern of scientific explanation – usually refer to the adaptational advantages (increasing fitness) related to particular mental modules (mechanisms). In short, on the basis of the evolutionary psychology, 'to explain something' means to show it's adaptive function. For example, typical evolutionary psychologist would say that cognitive mechanisms of face recognition have arisen as an adaption which enabled the recognition of relatives (which is crucial for kin selection – William Hamilton's inclusive fitness theory[34]) as well as recognition of the recipients of acts of altruism, from whom we expect reciprocation (which is crucial from the viewpoint of Robert Trivers' reciprocal altruism theory).[35]

In turn, the representatives of the embodied-embedded mind theory will definitely reject the postulates of the computability and modularity of level (3). Although they treat the theory of evolution seriously, they consider the above scheme to be naive (i.e. a scheme in which to explain means to show the evolutionary adaptive function). Not all of the products of evolution have an adaptive nature (most of the products of evolution are by-products).

The embodied-embedded mind paradigm is largely based on the achievements of such sciences as applied linguistics (especially the theory of conceptual metaphors by George Lakoff), anthropology (the theory of the cultural origins of human cognition by Michael Tomasello), and also on the new achievements of neurobiology (especially the theory of mirror neurons and embodied simulation).[36] Speaking

---

[33] See J.H. Barkow, L. Cosmides, J. Tooby, eds., *The Adapted Mind. Evolutionary Psychology and the Generation of Culture*, Oxford University Press, NY-Oxford 1992.

[34] See W.D. Hamilton, *The Genetical Evolution of Social Behaviour I and II*', "Journal of Theoretical Biology" 1964, vol. 7, pp. 1–16 and 17–32.

[35] See R. Trivers, *The Evolution of Reciprocal Altruism*, "The Quarterly Review of Biology" 1971, vol. 46, pp. 35–57, reprinted in *idem*, *Reciprocal Altruism*, [in:] *idem*, *Natural Selection and Social Theory. Selected Papers of Robert Trivers*, ed. S. Stich, Oxford University Press, NY 2008, pp. 3–55.

[36] Cf. G. Lakoff, *Women, Fire and Dangerous Things. What Categories Reveal about the Mind*, The University of Chicago Press, Chicago 1987; M. Tomasello, *The Cultural Origins of Human Cognition*, Harvard University Press, Cambridge 1999; V. Gallese, *Embodied Simulation*, "Phenomenology and Cognitive Sciences" 2005, vol. 4, pp. 23–48.

most generally, this paradigm shows an enormous role played by the physical interactions between individuals and the social and cultural environment in the shaping of cognitive abilities and mental states. Due to its openness to data from different branches of knowledge, the 'embodied-embedded mind' paradigm seems to be the most fertile of all the interpretative paradigms available. Nonetheless, it is not free of assumptions which are difficult to test empirically. An example of such an assumption is Lakoff's claim that the human conceptual system has a metaphoric nature. In the case of self-deception, the interpretative paradigms play a significant role, since particular neurocognitive models of self-deception are *explicite* or at least *implicite* based on different paradigms.

3. Three Neurocognitive Models of Self-Deception

It seems to us that – on the basis of contemporary cognitive neuroscience – one can distinguish at least three models of self-deception: (I) the computational model, (II) the evolutionary model and (III) the embodied mind model. We will show that these accounts are embedded in the following interpretational paradigms: computationalism (strong AI), evolutionary psychology and the embodied mind paradigm respectively. In the next part of this paper we will take the question of the biological conditions that an individual must satisfy in order to be able to deceive oneself. We shall argue that the necessary condition for self-deception is to pass a so-called false belief task, which is capable for individuals with a developed theory of mind. We will also try to indicate a connection between these models and the philosophical accounts of self-deception described in the first part of this paper.

3.1 The Computational Model

The computational model can also be called the Turing model. Of course, it should be pointed out that Turing was not concerned with the issue of self-deception or at least we do not know this to have been the case. To justify this nomenclature, let us briefly recall the famous Lucas-Penrose Gödelian Argument for the impossibility of creating an artificial intelligence by means of a computer algorithm.[37] According to this argument, the human mind is capable of performing certain mathematical operations which a computer cannot do. Strictly speaking, humans can ascribe a truth value to certain mathematical formulas and the Turing machine cannot. It is known that – according to Gödel's incompleteness theorem – a consistent formal system (that contains first-order arithmetic) cannot be complete. It has been proven that a consistent formal system is equivalent to a correct algorithm. One could undermine the Lucas-Penrose argument by stating that the human mind is an inconsistent formal system (the wrong algorithm) and therefore it can perform operations which are impossible on the grounds of consistent systems (correct algorithms). This solution has been taken into account by Alan Turing. In the lecture delivered in 1947 for the London Mathematical Society he noticed:

> In other words then, if a machine is expected to be infallible, it cannot also be intelligent. There are several theorems which say almost exactly that. But these theorems say nothing about how much intelligence may be displayed if a machine makes no pretence at infallibility.[38]

[37] Cf. J.R. Lucas, *Minds, Machines and Gödel*, "Philosophy" 1961, vol. 36, pp. 112–127; R. Penrose, *Shadows of the Mind. A Search for the Missing Science of Consciousness*, Vintage Books, London 2005.

[38] The quote is taken from R. Penrose, *Shadows of the Mind…*, *op. cit.*, p. 129.

Interestingly enough, a similar remark was made by Hilary Putnam in his private conversation with Lucas.[39] Computational efficiency could serve as an argument in favour of a modular theory of mind itself. It turns out that the calculations performed by systems with a modular architecture are more efficient than the calculations performed by centralized ones.

Jerry Fodor lists the following features of input systems which can be identified with modules: (i) input systems are domain specificity, (ii) the operation of input systems is mandatory, (iii) there is only limited central access to the mental representations that input systems compute (iv) input system are fast, (v) input systems are informationally encapsulated, (vi) input analyzers have 'shallow' outputs, (vii) input systems are associated with a fixed neural architecture, (viii) input systems exhibit characteristic and specific breakdown patterns (ix) the ontogeny of input systems exhibits a characteristic pace and sequencing. Since the modules are closed, insulated and specialized, it may be assumed that they are correct algorithms which are equivalent to a consistent formal system.[40]

The justification of an idea of "contradictory mind" can be found in the modular theory of mind.[41] This contradiction can manifest itself at the level of the integration of a greater number of modules which is required during more complex cognitive tasks. Enthusiasts of an interpersonal model of self-deception may attempt to embed it in a version of such modular theory. Those who decide to follow this way of thinking have many manoeuvres to deal with the paradoxes of self-deception at their disposal. It could be argued that in different men-

---

[39] "Professor Putnam has suggested that human beings are machines, but inconsistent machines. If a machine were wired to correspond to an inconsistent system, then there would be no well-formed formula which it could not produce as true; and so in no way could it be proved to be inferior to a human being", J.R. Lucas, *Minds, Machines and Gödel*, *op. cit.*, p. 121.

[40] See J.A. Fodor, *The Modularity of Mind*, *op. cit.*, pp. 47–101.

[41] One of the authors has already written about this in more depth in the following article: M.L. Hohol, *Umysł: system sprzeczny, ale nietrywialny*, "Zagadnienia Filozoficzne w Nauce" 2010, vol. 47, pp. 89–108.

tal modules, distinct representations of reality are created, of which only one is made aware. If so then a belief that *p* could be an effect of module X and a belief that not-*p* of a different module Y. Representation, which is made aware, could be inadequate, i.e., incompatible with reality. Generally speaking, there are many "familiar computational reasons for denying that an agent's beliefs are all inferentially integrated (the limitations of memory search strategies, etc.)".[42]

The problem is that the computational theory of mind (including its modular version) has been subject to withering criticism. As it has been already said, critiques were provided mainly by proponents of the 'embodied-embedded mind' paradigm. One can also find it controversial to identify mental representations and beliefs and claims that both of these notions require refinement. On the other hand, the advantage of the computational model is that it could give the answer to many questions concerning self-deception and shed some light on the mechanisms of this phenomenon. From this viewpoint, the contradictions cease to be puzzling and become rather a natural consequence of the functioning of the modular cognitive system.[43] In the history of science, however, we often deal with the theories that have a lot of explanatory power, although they have been proven to be false. Let us now turn to the evolutionary model.

3.2 The Evolutionary Model

In one of his books, Michael Gazzaniga writes: "To be a good liar, it helps not to know that you are lying or, in the case of psychopaths, not to care".[44] The evolutionary model of self-deception has been

[42] J.L. Bermudez, *Defending Intentionalist Accounts of Self-Deception*, "Behavioral and Brain Sciences", 1997, vol. 20, p. 107.
[43] A similar suggestion can be found in the following book: D. Livingstone Smith, *The Most Dangerous Animal: Human Nature and the Origins of War*, St. Martin's Press, NY 2007, chapter 6.
[44] M. Gazzaniga, *Human. The Science Behind What Makes Us Unique*, Ecco 2008, p. 102.

developed by Robert Trivers and his collaborators since 80's.[45] On the basis of this conception, self-deception is treated as an evolutionary adaptation which helps to deceive other people. Trivers summarizes his basic ideas in the following way:

> The central claim (...) is that self-deception evolves in the service of deception — the better to fool others. Sometimes it also benefits deception by saving on cognitive load during the act, and at times it also provides an easy defence against accusations of deception (namely, I was unconscious of my actions). In the first case, self-deceived fails to give off the cues that go with consciously mediated deception, thus escaping detection. In the second, the actual process of deception is rendered cognitively less expensive by keeping part of the truth in the unconscious. That is, the brain can act more efficiently when it is unaware of the ongoing contradiction. And in the third case, the deception, when detected, is more easily defended against—that is, rationalized—to others as being unconsciously propagated.[46]

In our opinion, in order to understand the evolutionary context in which Trivers' conception is placed, we ought to be aware of the so called *Machiavellian Intelligence Hypothesis* (MIH). Actually, it is not a single hypothesis but rather a set of several hypotheses and theories promoted by evolutionists and primatologists such as Nicholas Humprey, Frans de Wall and Andrew Whiten.[47] Despite the contro-

---

[45] Cf. R. Trivers, H.P. Newton, *The Crash of Flight 90: Doomed by Self-Deception?*, "Science Digest" 1982, vol. 111, pp. 66–67; R. Trivers, *Elements of a Scientific Theory of Self Deception*, "Annals of the New York Academy of Sciences" 2000, vol. 907, pp. 114–131; both reprinted in R. Trivers, *Self-Deception in Service of Deceit*, [in:] *Natural Selection and Social Theory..., op. cit.*, pp. 255–293; *idem*, *The Folly of Fools. The Logic of Deceit and Self-Deception in Human Life*, Basic Books, New York 2011, e-book edition; W. von Hippel, R. Trivers, *The Evolution and Psychology of Self-Deception*, "Behavioral and Brain Sciences" 2011, vol. 34, issue 1, pp. 1–56.

[46] R. Trivers, *The Folly of Fools..., op. cit.*, pp. 47–48.

[47] Cf. e.g. A. Whiten, *Machiavellian Intelligence Hypothesis*, [in:] *The MIT Encyclopedia of the Cognitive Sciences*, eds. R.A. Wilson, F.C. Keil, MIT Press, Mass. 1999, pp. 495–496.

versy that MIH often arouses, it is well established amongst the evolutionary sciences. By using MIH, evolutionary scientists try to explain for example the encephalization of the human brain, which started about 2 million years ago and was a milestone on the evolutionary way to *Homo sapiens.* For the purposes of this paper, a brief commentary will suffice. In its basic version, MIH speaks about the coevolution of cognitive abilities and the socialization of organisms. The 'Machiavellian' component states that socialization requires pragmatism ('Machiavellianism'). Climbing the ladder of a social hierarchy often requires lies, fraud, deception and the manipulation of other individuals. This in turn requires complex cognitive abilities. If somewhere there is a deceiver, a deceived is also present. This apparently trivial statement has, however, a deep evolutionary significance since the occurrence of cheaters and the cheated triggered an evolutionary arms race. As Trivers writes:

> Deceiver and deceived are trapped in a coevolutionary struggle that continually improves adaptations on both sides. One such adaptation is intelligence itself. The evidence is clear and overwhelming that both the detection of deception and often its propagation have been major forces favoring the evolution of intelligence. It is perhaps ironic that dishonesty has often been the file against which intellectual tools for truth have been sharpened.[48]

As we have said before, Trivers' model is strongly embedded in the paradigm of the evolutionary psychology. One of the basic assumptions of this paradigm says that human cognitive abilities have been formed in the so-called Environment of Evolutionary Adeptness (EEA) during the Pleistocene period. It is easy to imagine what penalties our ancestors might have born when they were caught deceiving. Those who decided to deceive had to be efficient in it and, sometimes, it was even a matter of life and death. According to this model,

---

[48] R. Trivers, *The Folly of Fools…*, *op. cit.*, p. 44.

self-deception allow us to cheat other people in an efficient way, since it minimizes the cognitive load of the deceiver. It reduces the risk of his exposure since, being self-deceived, he does not emit the physiological signals which are present during the conscious deception.

It is worth noting that it happens not only when a cheat is exposed to detection with an 'unaided eye'. Almost every method that forensics uses to detect a lie (from the ordinary polygraph through brain-fingerprinting, which is based on EEG writings, to sophisticated tests using fMRI) is based on an assumption that *a lie must be conscious.* If it is not, we cannot observe its physical correlates.

In Trivers' model, self-deception may also play other, less 'Machiavellian', functions such as self-enhancement, increasing our self-motivation or different psychological biases which consist of self-deception.[49] For instance, it occurs in human sexual strategies when men, seeking the favour of women (who – according to Trivers' parental investment theory – are the gender that makes the selection[50]) 'improve' their outlook or behaviour. David Buss points out that they very often deceive their potential female partners – in order to be perceived not as they really are, but in a way in which the potential female partners would like to perceive them.[51] This strategy, which helps to encourage the potential partner, of course, consists of self-deception. If Buss is right, it can be a serious argument for the adaptive nature of self-deception. He also postulates studies on the gender differentiation of deception and the self-deception strategies.

Trivers and von Hippel consider the 'nature' of self-deception and the cognitive mechanisms that allow for self-deception. They discard

---

[49] Cf. W. von Hippel, R. Trivers, *The Evolution and Psychology of Self-Deception*, *op. cit.*, pp. 12–15.

[50] Cf. R. Trivers, *Parental Investment and Sexual Selection*, [in:] *Sexual Selection and the Descent of Man 1871–1971*, ed. B. Campbell, Aldine Publishing Company, Chicago 1972, pp. 136–179, reprinted in R. Trivers, *Parental Investment and Reproductive success*, [in:] *Natural Selection and Social Theory...*, *op. cit.*, pp. 56–110.

[51] Cf. D.M. Buss, *The Evolution and Psychology of Self-Deception*, "Behavioral and Brain Sciences" 2011, vol. 34, issue 1, p. 18; *idem*, *The Evolution of Desire. Strategies of Human Mating*, Basic Books, NY 2003, *passim*.

the requirement that two separate and inconsistent representations of reality are stored in self-deceivers' minds (as we remember, such a view is present in the computational model). They write:

> Our approach of treating self-deception as information-processing biases that give priority to welcome over unwelcome information also differs from classic accounts that hold that self-deceiving individual must have two separate representations of reality, with truth preferentially stored in the unconscious mind and falsehood in the conscious mind (...). People can deceive themselves by preventing unwanted information from being encoded in the first place. (...) The individual need not have two representations of reality to self-deceive. Rather, people can self-deceive in the same way that they deceive others, by avoiding critical information and thereby not telling (themselves) the whole truth.[52]

Despite this, from the philosophical viewpoint the evolutionary approach to self-deception still seems to be more consistent with the intentionalist models based on interpersonal deception. Although it rejects the requirement of holding of contradictory beliefs, the theory retains the idea of split selves. According to von Hippel and Trivers, self-deception involves the dissociations of conscious and unconscious cognitive processes, including: implicit versus explicit memory, implicit versus explicit attitudes and automatic versus controlled processes. These mechanisms "ensure that the mental processes that are the target of self-deception do not have access to the same information as the mental processes deceiving the self".[53]

On the other hand, the Authors argue that these mechanisms are biased in a way that shows self-deceivers' goals and motivations, which could make their approach similar to that of Mele. This claim is made *explicite* in their article of 2011. We believe, however, that their theory is more akin to the conceptualization put forth by Davidson.

---

[52] *Ibid.*, p. 2.
[53] *Ibid.*, p. 6.

As we have pointed out, the main difference between these two philosophical approaches to desire-influenced biasing is that, in the latter, the self-deceiver must be aware of the evidence against a belief that she wants to be true. Von Hippel and Trivers suggest that the flexibility of information-gathering bias allows for such awareness.[54]

The main advantage of the Trivers' conception lies in showing the adaptive role of self-deception. The problem is that Trivers *implicite* assumes that people are able to detect the lies and deception of others efficiently. Obviously, this assumption can be subjected to empirical testing.[55] This task was undertaken by Paul Ekman, who showed that particular emotions have their correlates in facial expressions.[56] Mimicry is therefore one of the most important indicators in exposing liars. Ekman's research showed that human cognitive abilities in detecting lies are still only slightly developed. Obviously, there are exceptions to this rule (in Ekman's opinion, among them are, for example, agents of secret services and psychotherapists), but, in general, the results of studies are unambiguous. From a sample of 12, 000 people, only 20 showed above-average capabilities in this area. A liar usually experiences strong feelings such as fear. The problem with detecting lies on the basis of the mimicry consists, inter alia, of the fact that sources of others' feelings are not always interpreted properly.[57]

Moreover, there is no general agreement as to the possibility of expanding the concept of self–deception beyond the evolutionary arms race, on phenomena like self-enhancement or self-motivation. For example, Steven Pinker – who is a strong supporter of the adaptive interpretation of the self-deception as well – claims that the extrapolation of this phenomenon over other phenomena may be misleading.[58]

---

[54] Cf. *ibid.*, p. 2.

[55] Cf. D. Dunning, *Get Thee to a Laboratory*, "Behavioral and Brain Sciences" 2011, vol. 34, issue 1, pp. 18–19.

[56] Cf. P. Ekman, *Telling Lies: Clues to Deceit in the Marketplace, Politics, and Marriage*, W.W. Norton & Company, NY 1985, *passim*.

[57] Cf. M. Gazzaniga, *Human…*, *op. cit.*, pp. 103–104.

[58] Cf. S. Pinker, *Representations and Decision Rules in the Theory of Self-Deception*, "Behavioral and Brain Sciences" 2011, vol. 34, issue 1, p. 36.

Finally, some scientists completely undermine the adaptive nature of self-deception. In this context, Hugo Mercier says that "most of the results used by the authors [von Hippel and Trivers – M.L.H., P.U.] can be accounted for as a by-product of a confirmation bias inherent in reasoning that does not have a self-deceptive function".[59] Perhaps Trivers himself became a victim of self-deception in his studies on the nature of the self-deception?

When the matter concerns cognitive mechanisms that enable self–deception, the evolutionary model does not reach beyond level (3). The next model we describe combines data from levels (1), (2) and (3) in a much better fashion.

3.3 The 'Embodied Mind' Model

In our opinion, the 'embodied mind' model is composed of two different, but in many ways complementary, theories of self-deception. The first of them was formulated by Vilyanur S. Ramachandran, and the second by William Hirstein (who was a student of the former). We include those theories in the embodied model for at least two reasons which we will now examine in more depth.

While the starting point for the two previous models was self-deception among 'healthy' people, the starting point of the Ramachandran conception were studies on patients with anosognosia.[60] Anosognosia is a clinical disorder which manifests itself in an inability to see one's own disease or denying its very existence. Usually it is – to varying degrees – accompanied by such phenomena as rationalization, false memory formation or 'phantom limb' delusions and the like. Disorders are most severe in lesions of the right cerebral hemisphere

[59] Cf. H. Mercier, *Self-Deception: Adaptation or By-product?*, "Behavioral and Brain Sciences" 2011, vol. 34, issue 1, p. 35.

[60] See V.S. Ramachandran, *The Evolutionary Biology of Self-Deception, Laughter, Dreaming and Depression: Some Clues from Anosognosia*, "Medical Hypotheses" 1996, vol. 47, pp. 347–362.

(especially right parietal lobe). It should be added that in many cases anosognosia is accompanied by complete omission of the left (i.e., contralateral) side of the body, which is called hemineglect.[61] A conversation below is a typical example of anosognosia. It was held between Ramachandran (initials VSR) and a female patient following a right hemisphere stroke (initials FD). FD was in a wheelchair and she could not move her left arm:

> **VSR**: Mrs D, how are you feeling today? **FD**: I've got a headache. You know, doctor, I've had a stroke so they brought me to the hospital.
> **VSR**: Mrs D, can you walk? **FD**: Yes. (*FD had been in a wheelchair for the past two weeks. She cannot walk.*)
> **VSR**: Mrs D, hold out your hands. Can you move your hands? **FD**: Yes.
> **VSR**: Can you use your right hand? **FD**: Yes.
> **VSR**: Can you use your left hand? **FD**: Yes.
> **VSR**: Are both hands equally strong? **FD**: Yes, of course they are equally strong (…).
> **VSR**: Can you point to my nose with your right hand? **FD**: (*She followed the instructions and pointed to my nose.*)
> **VSR**: Mrs D, point to me with your left hand. **FD**: (*Her hand lay paralyzed in front of her.*)
> **VSR**: Mrs D, are you pointing my nose? **FD**: Yes.
> **VSR**: Can you clearly see it pointing? **FD**: Yes, it is about two inches from your nose (…).
> **VSR**: Mrs D, can you clap? **FD**: Of course I can clap.
> **VSR**: Mrs D, will you clap for me? **FD**: (*She proceeded to make clapping movements with her right hand as if clapping with an imaginary hand near the midline!*)
> **VSR**: Are you clapping? **FD**: Yes, I'm clapping (…).[62]

[61] See K.W. Walsh, D. Darby, *Neuropsychology. A Clinical Approach*, 5 ed., Elsevier, NY 2005, chapters 3 and 6.

[62] V.S. Ramachandran, *The Evolutionary Biology of Self-Deception…*, *op. cit.*, pp. 348–349.

Ramachandran claims that in her case we are dealing with self-deception. Mrs. FD not only does not acknowledge the truth about her illness, but also creates a fictional representation of her body and reality. Why does this happen? Anosognosia may be explained in psychodynamic categories.[63] In such an approach, neglecting the existence of one's own illness would be a psychological defence mechanism – a reaction against suffering.

Ramachandran rejects such an interpretation, since agnosnosia is less frequently observed as a result of the lesion of the left cerebral hemisphere.[64] It serves as a clue that the neuronal mechanisms of the anosognosia can be found in the right cerebral hemisphere, not in the mental. Ramachandran puts his own hypothesis about the formation of self-deception and other defence mechanisms:

> The real reason for the evolution of these defense mechanisms (confabulations, rationalization), I suggest, is to create a coherent belief system in order to impose stability in one's behavior.[65]

He claims that this hypothesis is strictly connected with hemispheric specialization. According to the commonly accepted, yet very imprecise, theorem, the left hemisphere is specialized in language, and the right is specialized in visual and spatial tasks as well as those tasks which require creativity. Moreover, the well-known experiments conducted by Michael Gazzinga on patients with split-brains suggest that the left hemisphere also plays the role of the interpreter of reality.[66] The objective of this left-brain interpreter is to organize the available data, even if the representation created is not compatible with the reality.

---

[63] See e.g. H. Grzegołowska-Klarkowska, *Samoobrona przez samooszukiwanie się*, [in:] *Złudzenia, które pozwalają żyć*, eds. M. Kofta, T. Szustrowa, PWN, Warszawa 2001, pp. 176–198.

[64] See V.S. Ramachandran, *The Evolutionary Biology of Self-Deception…*, *op. cit.*, p. 350.

[65] *Ibid.*, p. 351.

[66] Cf. e.g. M. Gazzaniga, *Human…*, *op. cit.*, part 3, chapter 8.

The results of experiments conducted by Gazzinga also show that the left-brain interpreter rationalizes the reality *ex post*, e.g. causing the creation of false information about the motivations of one's own actions. In his conception of hemisphere specialization, Ramachandran goes a step further:

> The basic idea here is that the *coping strategies* of two hemispheres are fundamentally different. The left hemisphere's job is to create a model and maintain it at all costs. If confronted with some new information that doesn't fit the model, it relies on Freudian defense mechanisms to deny, repress or confabulate; anything to preserve status quo. The right hemisphere's strategy, on the other hand, is fundamentally different. I like to call it the 'anomaly detector', for when the anomalous information reaches a certain threshold, the right hemisphere decides that it is time to force the left hemisphere to revise the entire model and start from scratch.[67]

In the 'embodied mind' model, mechanisms of self-deception are connected with the functioning of the left-brain interpreter. The imposition of consistency on a belief system at all costs results in the forming of false representations concerning reality. The functioning of the interpreter favours the 'cognitive economy' and it allows fast and efficient decision making. A consistent picture is formed on the basis of, for example, the selection of relevant information, the categorization of information with respect to its truthfulness and falsity, control of initiated actions, and prediction of their consequences.[68] When some kind of threshold of 'incompatibility with reality' is exceeded, the right cerebral hemisphere induces the left-interpreter to acknowledging the new data and conduct a revision of the belief system.

---

[67] V.S. Ramachandran, *The Evolutionary Biology of Self-Deception…*, *op. cit.*, p. 352.
[68] See, A. Herzyk, *Neuropsychologia kliniczna wobec zjawisk świadomości i nieświadomości*, PWN, Warszawa 2012, p. 137.

Although self-deception is – to some extent – normal among healthy people, in the case of left-hemisphere lesion patients it takes a pathological form.[69] Ramachandran explains it in the following way:

> In patients suffering from anosognosia, the left hemisphere is doing all of the confabulation and denial as it would in a normal person. The difference is that these patients have lost the mechanism in the right hemisphere that would ordinarily force them to generate a paradigm shift in response to conflicting information. This forces the patient into a delusional trap and he will continue to confabulate without switching paradigms. The patient may therefore glibly explain away any anomaly or discrepancy so that he is, on the whole, blissfully oblivious to his dire predicament.[70]

In the 'embodied mind' model, self-deception is made possible by making various data cohesive, where the emphasis is not placed on adequateness, but on the consistency of the data. Hence, this model has little in common with the approach to self-deception that requires maintaining of contradictory beliefs. The mechanism of self-deception consists rather of filtering, blocking and ignoring information that is inconsistent with the belief system.

Studies on phantom limb pain and Capgras syndrome could also shed some light on mechanisms of self-deception. In the first case, the patient is convinced about pain felt in the amputated limb. It means that he has a false representation of his own body.[71] It is probable that the mechanism of self-deception is therefore very similar to the mechanism of agnosnosia described by Ramachandran. In the case of Capgras syndrome, the matter becomes more complicated since

---

[69] To the point of this distinction see e.g. N. Levy, *Neuroethics*, Cambridge University Press, Cambridge-NY 2007, chapter 8: *Self-Deception: the Normal and the Pathological*, pp. 258–280.

[70] V.S. Ramachandran, *The Evolutionary Biology of Self-Deception…*, *op. cit.*, p. 352.

[71] See e.g. V.S. Ramachandran, S. Blakeslee, *Phantoms in the Brain. Probing the Mysteries of the Human Mind*, Quill William Morrow, NY 1998.

Capgras patients create false beliefs about their relatives.[72] Typically, they claim that their relatives were replaced by understudies who pretend to be them, look like them and behave in the same way. Due to incoherence in the behaviour of people affected with Caprgas syndrome, this issue may be interesting for researchers of self-deception.

One neuronal hypothesis connects Caprgas syndrome with dysfunction in the frontal cortex. The representation of the relative's identity is maintained, but mistakenly interpreted due to disorders of reality monitoring mechanisms. The dissociation concerns the *affective state* controlled by the limbic system and the processing of visual information, in which a crucial role is played by the temporal areas.[73] The improper functioning of the 'emotional brain' also plays a significant role. Acquiring a false belief about the 'replacement' of relatives by 'understudies' may be an effect of a lack of an emotional reaction to the faces of those relatives.

The second theory of the 'embodied mind' model is proposed by William Hirstein in his book *Brain Fiction.*[74] Like Ramachandran, he starts from clinical data and then analyses self-deception on many levels (neuronal structures, representations). At the level of neuronal structures, Hirstein argues that mechanisms involved in self-deception are the same as those which play a significant role in pathologic confabulation and Obsessive-Compulsive Disorder. Above all, Hirsetein emphasizes two facts. Firstly:

> (…) damage to the mediodorsal nucleus of the thalamus is often present in confabulating patients. Anatomists Ray and Price note that the thalamic mediodorsal nucleus may function to block cortical activity: "If thalamocortical activity provides a mechanism for sustaining activity related to concentration on a particular task, inhibition of [the

[72] See K.W. Walsh, D. Darby, *Neuropsychology. A Clinical Approach*, 5 ed., Elsevier, NY 2005, chapter 11.

[73] See, A. Herzyk, *Neuropsychologia kliniczna…, op. cit.*, pp. 153–155.

[74] See W. Hirstein, *Brain Fiction. Self-Deception and the Riddle of Confabulation*, MIT Press, Cambridge-London 2005.

> mediodorsal nucleus] may block sustained activity, in order to switch between different patterns of thalamocortical activity, or to suppress unwanted thoughts or emotions".[75]

According to Hirstein, in cases of the hyperactivity of this neuronal structure, we are dealing with self-deception. This conclusion is consistent with the general idea of Ramachandran, who argues that self-deception consists of the inability to incorporate new incoherent content into the consistent belief system. Secondly, Hirstein draws attention to the orbitofrontal lesions that result in the handicapping of the inhibitory emotions:

> There is also an emotional component involved in sustaining self--deceptive beliefs and beliefs that give rise to confabulations. (...). The failure of their brains to generate the proper inhibitory emotions seemed to make these patients behave in disinhibited ways. Emotions may also play a role in getting self-deceived people to give up a self--deceptive belief permanently. Without the proper emotional response, the person might admit his evidence is weak, but then shortly thereafter reaffirm self-deceptive belief, just as confabulators do. In many cases, self-deceived person avoids thinking a crucial thought with an emotional feeling of conviction, in the way that sociopaths do not feel emotions deeply. Sociopaths who are capable of thinking a crucial thought (I should not be hurting people like this) with conviction are self-deceived; those who are not capable of such conviction are self-deceived in a tension-free way. Even in normal people, to be self-deceived is not to never think a feared thought. No one can avoid thinking such a thought at least sometimes, in the same way that one cannot avoid thinking of camels once ordered not to. But one can think a thought without conviction, especially if one's feelings of conviction are weak or disorganized.[76]

---

[75] *Ibid.*, p. 228.

[76] *Ibid.*

If the matter concerns a level higher than the neuronal, namely, the representation level, Hirstein simply rejects the possibility of having two contradictory beliefs and formulates arguments similar to those of Mele (to which he *explicite* refers). According to Hirstein, the same information may be represented in a different way, either in the formal, linguistic or neuronal form. In cases of self-deception, the contradiction in the strictest sense occurs not between two beliefs, but between representations of the same information which are located on completely different cognitive levels. As he writes:

> A person can also represent something as being the case without having a belief that it is the case. In addition to beliefs, the brain also contains topographic representations, which are a variety of analog representation. One source of difficulty may be that we have trouble thinking of the information contained in the brain's topographic, or analog representations as consisting of beliefs (…). An analog representation can be in conflict with a belief in the same way that a picture and a sentence can be in conflict. This suggests a way to resolve a question about self-deception. How can a person selectively avoid evidence for the belief that *p* without in some way believing that *p*? How else does he know what evidence to avoid? One way to deal with this is as follows. The information that *p* is already represented in analog form. There is conflict in the person's mind, but the conflicting information is represented in two different forms, conceptual and analog. This is different from holding two contradictory beliefs in full conceptual form. What may be happening is that the brain has a way of preventing certain types of analog information from being represented in conceptual form, from being explicitly thought and believed.[77]

Summarizing the 'embodied mind' model, we can agree with Ramachandran that his theory is highly speculative and formulated in

---

[77] *Ibid.*, p. 229.

some sort of metonymic terms. An argument for this conception is the coherence of the various data coming from studies on lesions as well as from experiments – both clinical and behavioural – which have been designed and conducted by Ramachandran.[78] Although his theory is interesting and coherent, we believe that it cannot be considered as the final explanation of the phenomenon of self-deception. Basically, we can say the same about Hirstein's conception. Nevertheless, both theories taken together probably give the best possible explanation of the mechanisms of self-deception which we have at present.

Both are also hard to collate with philosophical models. Although we hesitate to indicate some strong connections, we are inclined to argue that they still have more in common with philosophical models based on interpersonal deception. It could be claimed that the idea of the left-brain interpreter that rejects the inconsistent data perceived by the right-brain militate in favour of split-selves approaches. One who is amazed by split-brain experiments might be tempted to argue that both hemispheres have a different physical basis for the two separate subsystems of the mind. One can also consider the conflict between the belief that *p* and the same information represented in other form as a conflict between two such mental subsystems. We believe that this way of thinking should be regarded as an oversimplification. Although the matter seems to be more sophisticated, it is easy to see that intentionalist approaches can be held as more coherent with the theories of Ramachandran and Hirstein.

4. Self-deception and the Theory of Mind

Neurocognitive models of self-deception need to be supplemented with some remarks on the Theory of Mind. This matter is relevant since to a large extent it will decide whether self-deception will be

[78] Cf. e.g., *ibid.*, p. 353–356; *idem*, *Anosognosia in Parietal Lobe Syndrome*, "Consciousness and Cognition" 1995, vol. 4, pp. 22–51.

considered as human-specific or as a phenomenon that occurs also among other animals. It can be argued that to be able to self-deceive one first needs to be capable of ordinary deception. Although tactical deception is common in wildlife, research conducted by Michael Tomasello and his team showed that only primates can intentionally deceive.[79] We believe, however, that this condition should be strengthened and the ability to deceive correlates with success in passing the false-belief task. This empirical experiment examines the capacities for ascribing a false belief to others and predicting their behaviour on the basis of their assumed mental states. One of the versions of a false-belief task appears as follows:

> (...) children watch as a treat is hidden in a specific location (e.g. in a box). Another person (Maxi) is present when the treat is hidden but then leaves the room, at which time the treat is moved to a new location as the children watch. The children are then asked where Maxi will look for the treat when he returns. A robust finding is that most 4 year-old children can solve the problem, stating that Maxi will look where the treat was originally hidden, whereas most younger children state that Maxi will look for the treat in the new hiding place, apparently not realizing that Maxi' s knowledge is different from their own.[80]

So, empirical research shows that four year old children are able to attribute a belief that they know to be false. To succeed in a false-belief task one must have one's own Theory of Mind developed.[81] The process of forming the Theory of Mind begins with the so-called nine-

---

[79] Cf. B. Hare, J. Call, M. Tomasello, *Chimpanzees Deceive a Human by Hiding*, „Cognition" 2006, vol. 101, pp. 495–514.

[80] D.F. Bjorklund, J.M. Bering, *Big Brains, Slow Development and Social Complexity: the Developmental and Evolutionary Origins of Social Cognition*, [in:] *Social Brain: Evolution and Pathology*, eds. M. Brüne, H. Ribbert, W. Schiefenhövel, Wiley, Chichester 2003, p. 127.

[81] Cf. S. Baron-Cohen, *Mindblindness*, MIT Press, Boston 1997, especially chapter 4: *Developing Mindreading: Four Steps*, pp. 31–58.

-month revolution, when children start to treat other people as intentional beings, namely, beings capable of achieving their own goals.[82] Under normal circumstances, the process of obtaining the Theory of Mind ends at the age of four. Then children are capable of 'reading in other humans' minds, i.e., creating simulations of their beliefs, needs and feelings. Generally speaking, Theory of Mind allows us to resign from our first person perspective and to adapt to the perspective of someone else. If an ability to deceive is a condition *sine qua non* for self-deception, it seems that having this competence is crucial. Michael Tomasello claims that non-human primates cannot pass the false-belief task (its non-verbal version).[83] Hence, if Tomasello and ourselves are right, the only individuals capable of self-deception are humans.

Conclusion

In this paper we have presented and discussed three neurocognitive models of self-deception: the computational, the evolutionary and the 'embodied mind' model. This presentation was preceded by an overview of the approaches that have emerged from the philosophical debates on the topic. Those approaches, in turn, can be organized into two main groups – the intentionalist and anti-intentionalist. Several conclusions can be drawn from this juxtaposition.

First of all, one can hold the conviction that the general disagreement among the philosophers effectively ruins their project. The authors of the first empirical research about the functions and mechanisms of self-deception – Gur and Sackeim – called their well-known paper *Self-Deception: A Concept in Search of a Phenomenon*. Indeed, looking at the philosophical disputes, one can come to the conclusion that the search is still on. "Those who invoke the word self-deception

[82] See M. Tomasello, *The Cultural Origins of Human Cognition*, *op. cit.*, pp. 61–77.

[83] See *ibid.*, p. 176.

to represent one phenomenon often argue that those who use it to represent another are misusing the construct".[84] Although there is still much to be done in order to mediate between these approaches, we have to admit that philosophy has provided particularly valuable insights concerning the issue.

Secondly, one can argue that the resolution of the paradoxes of self-deception will stem from neurological theories. Although this account is close to one we would take, in this respect we are far from excessive optimism. Each empirical study should carefully consider which conception of self-deception it is utilizing. Even if one finally admits that self-deception is a fuzzy concept and that neurological evidence about its nature mediates its various forms,[85] the philosophical burden on the assumptions underlying the theories is still in question. Furthermore, we are still far from a good neuroscientific model of self-deception. As Dennis Krebs, J'Anne Ward, and Tim Racine put it:

> We need to determine what, exactly, a belief is; how people form beliefs, and where, and in what forms, they exist in the brain. We need to develop better models of the self – the knower, or information processor – alleged to be the agent and object of deception. We need to learn more about how information is stored, differentiated, and integrated in the brain and the extent to which independent neural structures process information in parallel ways.[86]

And this theoretical work is to be done by neuroscientists and neuroscientifically oriented philosophers as well.

Thirdly, there remains the question of the correspondence between the philosophical approaches to self-deception and scientific ones. It seems to us that intentionalist approaches based on interper-

---

[84] D. Krebs, J. Ward, T. Racine, *The Many Faces of Self-Deception*, "Behavioral and Brain Sciences", 1997, vol. 20, p. 119.

[85] Cf. *ibid.*

[86] *Ibid.*

sonal deception can be considered as more coherent with the models of cognitive neuroscience presented in this paper. We believe, however, that the conceptualization proposed by the anti-intentionalists may be more useful in other empirical investigations on the topic (e.g. psychological ones).

And, last but not least, we believe that self-deception is a typical issue of philosophy *in* science. It is a traditional philosophical problem intertwined with empirical theories which often emerge from the conceptual analyses undertaken by philosophers. In reflections over self-deception, one can implement the achievements of contemporary neuroscience, but then one should be careful about some of the assumptions of the theories used. We believe that the question of self-deception should be taken into consideration in the research program of philosophy *in* science, to which this study constitutes a humble contribution.

Jerzy Vetulani
Polish Academy of Sciences
Copernicus Center for Interdisciplinary Studies

The Neurobiology of Depression[1]

Introduction

Depression is one of the most frequent psychiatric conditions. Every third human being experiences feelings of depression at least once during his or her lifetime, and at least one sixth of the population has encountered at least one episode of major depression. Moreover, epidemiological studies indicate that the frequency of depression increases.[2] It is speculated that this is a result of increasing social stresses and decreasing social support (for instance from the family) due to changes in social lifestyle, and those changes mostly affect young persons. The trend of increasing unemployment among young men and the increase in the number of working women may be responsible for the fact that, while depression previously dominated among women,[3]

---

[1] This paper was written within the research grant "The Limits of Scientific Explanation" sponsored by the John Templeton Foundation. It was originally published in Polish as an introductory chapter in a book "Farmakoterapia depresji", edited by J. Heitzman and J. Vetulani, published by Termedia in 2012, and is published here by permission of Termedia.

[2] E. Fombonne, *Increased Rates of Depression: Update of Epidemiological Findings and Analytical Problems*, "Acta Psychiatrica Scandinavica" 1994, vol. 90, pp. 145–156.

[3] R.C. Kessler, K.A. McGonagle, M. Swartz, D.G. Blazer, C.B. Nelson, *Sex and Depression in the National Comorbidity Survey I: Lifetime Prevalence, Chronicity and Recurrence*, "Journal of Affective Disorders" 1993, vol. 29, pp. 85–96.

presently the morbidity within both sexes is becoming similar,[4] and the suicide rate among young men is increasing.[5]

The history of the progress of our knowledge about the neurobiology of depression was, at the beginning, closely associated with the progress of the pharmacotherapy of depression, and whatever bad can be said about the use of antidepressants, it is due to the steady progress in this field that the number of cases requiring hospitalization has declined, the periods of hospitalization have shrunk, and periods of remission stretch to such an extent that properly treated patients with depression may regularly participate in social life.

Although known since antiquity – Helen of Troy served a magic potion νεπεντηεσ (a mixture of poppy and wine) to alleviate soldiers' sorrow and grief – depression was for a long time ignored by medicine, as it was rather harmless for the patient's environment. Depression was regarded as "vapours", and, in 1763, Pierre Pomme, the general practitioner to the French King Louis XV, claimed to have discovered its cause: "These nervous vapours, arising as they did from the uterus" and added that they could "derange all functions of the brain". He proposed to cure it with chicken soup and cold baths.[6] In the 19th century, depression, together with hypochondria and hysteria, was classified as a "neurotic disease".

The progress in knowledge of neurobiological disturbances underlying depression began in the middle of the 20th century, when it was demonstrated that a potent antipsychotic and antihypertensive drug, reserpine, introduced at that time, caused strong symptoms of depression and at the same time drastically depressed the brain levels of biogenic amines – norepinephrine and serotonin.[7] To complement

---

[4] M.M. Weissman, R. Bland, P.R. Joyce, S. Newman, J.E. Wells, H.U. Wittchen, *Sex Differences in Rates of Depression: Cross-national Perspectives,* "Journal of Affective Disorders" 1993, vol. 29, pp. 77–84.

[5] D.J. Pounder, *Changing Patterns of Male Suicide in Scotland*, "Forensic Science International" 1991, vol. 51, pp. 79–87.

[6] P. Pomme, *Traite des Affections Vaporeuses des Deux Sexes*, Lyon 1767.

[7] P.A. Shore, S.L. Silver, B.B. Brodie, *Interaction of Reserpine, Serotonin, and Lysergic Acid Diethylamide in Brain*, "Science" 1955, vol. 122, pp. 284–285.

this finding it was discovered that the first antidepressants introduced at that time were inhibitors of monoamine oxidase (MAO), and they effectively elevated the cerebral concentrations of biogenic amines.[8] Almost at the same time a novel antidepressant – imipramine – was discovered:[9] the drug did not affect monoamine levels, but it was quickly discovered that it inhibited the reuptake of released norepinephrine and serotonin to neurons, from which they were released, and in this manner increased their concentration in the synaptic cleft, thus potentiating their interaction with specific receptors.[10]

Despite intensive research on major depression during the last fifty years, the detailed neurobiological processes and molecular mechanisms underlying this heterogeneous disease still remain poorly understood, although, paradoxically, several rather effective new therapies have been developed. Aside from new efficient drugs, one should mention the progress made in psychotherapy (particularly cognitive-behavioural), seismotherapy (electroconvulsive shock and transcranial magnetic stimulation), phototherapy, REM sleep deprivation, and deep brain stimulation.

With every passing decade, our knowledge about depression grows considerably, owing to introduction of new biochemical methods, progress in neuroimaging, successes of neurosurgery and developments in research on genetic basis of depression. We now know much more than thirty years ago, but it does not mean that we can see the problem more clearly, as the initially simple picture has become greatly complicated.

---

[8] H.P. Loomer, J.C. Saunders, N.S. Kline, *A Clinical and Pharmacodynamic Evaluation of Iproniazid as a Psychic Energizer*, "Psychiatric Research Reports. American Psychiatric Association" 1957, vol. 8, pp. 129–141.

[9] R.Kuhn, *The Treatment of Depressive States with G 22355 (imipramine hydrochloride)*, "American Journal of Psychiatry" 1958, vol. 115, pp. 459–464.

[10] J. Axelrod, L.G. Whitby, G. Hertting, *Effect of Psychotropic Drugs on the Uptake of H3-norepinephrine by Tissues*, "Science" 1961, vol. 133, pp. 383–384.

1. Monoamines and Depression

Owing both to historical aspects and the high position of antidepressants that are selective serotonin reuptake inhibitors (SSRI's), the position of monoamines in the neurobiological theories of depression is still strong. The experimental proof of that is that both depressive patients in the remission period and healthy persons prone to depression exhibit depressive symptoms after being given a diet deficient in tryptophan[11], an amino acid that is a precursor of serotonin.[12]

Since the beginning of studies on the role of monoamines in the etiology of depression, the dispute over which one played a key role has taken place. In the 1960's there was a contest between the advocates of "noradrenergic"[13] and "serotonergic"[14] theory. Neither of those hypotheses explained all the aspects of antidepressant action and the etiology of depression, and therefore the attention had been focused on the third monoamine, the latest recognized as a neurotransmitter, namely dopamine. Dopamine is strongly associated with the activity of the reward system. The impairment of this system leads to amotivation and anhedonia, the phenomena present in depression. The "dopaminergic" hypothesis of depression was proposed by Axel Randrup[15, 16]

---

[11] P.L. Delgado, D.S. Charney, L.H. Price, G.K. Aghajanian, H. Landis, G.R. Heninger, *Serotonin Function and the Mechanism of Antidepressant Action. Reversal of Antidepressant-induced Remission by Rapid Depletion of Plasma Tryptophan*, "Archives of General Psychiatry" 1990, vol. 47, pp. 411–418.

[12] C. Benkelfat, M.A. Ellenbogen, P. Dean, R.M. Palmour, S.N. Young, *Mood-lowering Effect of Tryptophan Depletion. Enhanced Susceptibility in Young Men at Genetic Risk for Major Affective Disorders*, "Archives of General Psychiatry" 1994, vol. 51, pp. 687–697.

[13] J.J. Schildkraut, *The Catecholamine Hypothesis of Affective Disorders: a Review of Supporting Evidence*, "American Journal of Psychiatry" 1965, vol. 122, pp. 509–522.

[14] A. Coppen, *The Biochemistry of Affective Disorders*, "British Journal of Psychiatry" 1967, vol. 113, pp. 1237–1264.

[15] A. Randrup, E. Mogilnicka, *Spectrum of Pharmacological Actions on Brain Dopamine. Indications for Development of New Psychoactive Drugs. Discussion of Amantadines as Examples of New Drugs with Special Actions on Dopamine Systems*, "Polish Journal of Pharmacology & Pharmacy" 1976, vol. 28, pp. 551–556.

[16] A. Randrup, C. Braestrup, *Uptake Inhibition of Biogenic Amines by Newer Antidepressant Drugs: Relevance to the Dopamine Hypothesis of Depression*, "Psychopharmacology (Berl)" 1977, vol. 53, pp. 309–314.

and Jerzy Maj[17], and was corroborated by the finding that the depression of activity of dopaminergic system is associated with psychomotor retardation, characteristic for depression,[18] and functional changes during remission are observed in such dopamine-rich brain areas, as basal ganglia and medial prefrontal cortex.[19]

It seems that the amphetamine-like effect of REM sleep deprivation[20] is responsible for the rather high antidepressant efficiency (60% improvement) of this procedure.[21] Also some novel antidepressants evidently stimulate the dopaminergic system of the brain. The best known of those is bupropion, a dopamine and norepinephrine reuptake inhibitor,[22] a compound of two identities: used as Welbutrin as an antidepressant and as Zyban to reduce the severity of nicotine craving and withdrawal symptoms, thus helping smoking cessation.

An interesting attempt to find out which one of the three main monoamines plays the critical role in depression and antidepressant action was a recent study on transgenic mice lacking the transporter of one of those monoamines, It was speculated that if a depression of neurotransmission related to one of the amines would cause depression, a strain lacking the transporter for this particular amine, and

---

[17] J. Maj, *Central Effects Following Repeated Treatment with Antidepressant Drugs*, "Polish Journal of Pharmacology & Pharmacy" 1984, vol. 36, pp. 87–99.

[18] P.J. Shah, A.D. Ogilvie, G.M. Goodwin, K.P. Ebmeier, *Clinical and Psychometric Correlates of Dopamine D2 Binding in Depression*, "Psychological Medicine" 1997, vol. 27, pp. 1247–1256.

[19] G.M. Goodwin, M.P. Austin, N. Dougall, M. Ross, C. Murray, R.E. O'Carroll, A. Moffoot, N. Prentice, K.P. Ebmeier, *State Changes in Brain Activity Shown by the Uptake of 99mTc-exametazime With Single Photon Emission Tomography in Major Depression Before and After Treatment*, "Journal of Affective Disorders" 1993, vol. 29, pp. 243–253.

[20] D. Ebert, *Sleep Deprivation and Dopamine – the Psychostimulant Theory of Antidepressant Sleep Deprivation*, [in:] *New Models for Depression*, eds. D. Ebert, K.P. Ebmeier, Karger, Basel 1998, pp. 153–169.

[21] J.C. Wu, W.E. Bunney, *The Biological Basis of an Antidepressant Response to Sleep Deprivation and Relapse: Review and Hypothesis*, "American Journal of Psychiatry" 1990, vol. 147, pp. 14–21.

[22] S.M. Stahl, J.F. Pradko, B.R. Haight, J.G. Modell, C.B. Rockett, S. Learned-Coughlin, *A Review of the Neuropharmacology of Bupropion, a Dual Norepinephrine and Dopamine Reuptake Inhibitor*, "Primary Care Companion to The Journal of Clinical Psychiatry" 2004, vol. 6, pp. 159–166.

thus unable to reduce rapidly its concentration in the synaptic cleft, would behave as wild-type mice treated with an inhibitor of a given transporter and display an "antidepressant" type of behaviour. A comparison of strains lacking the transporter of serotonin (SERT), dopamine (DAT) and norepinephrine (NAT) has proven, to some surprise, that only the latter mice display "antidepressant" behaviour. The animals lacking SERT displayed higher timidity, but not "antidepressant" behaviour, while DAT knockouts were characterized primarily by abnormally high motility, turning into stereotyped behaviour. In the tests for "antidepressant" effect they showed some activity, though much less expressed than in NAT knockouts.[23] Thus, it seems that after a long period of domination of the view about the crucial role of serotonin (basing on the efficacy of SSRI's) there is a return to the noradrenergic hypothesis of Schildkraut, formulated in the mid-60's.[24]

Monoaminergic hypotheses, according to which depression is a neurochemical disease resulting from the hypoactivity of noradrenergic, dopaminergic or serotonergic neurotransmission, dominated the first four decades of studies on neurobiology of depression, but began to meet more and more difficulties.

One of the first problems encountered was that of therapeutic delay. Although the neurochemical effects of drugs affecting neurotransmission developed rapidly, within minutes or hours monoamine oxidase or monoamine transporters were blocked, the clinical improvement appeared only after few weeks of regular administration of drugs.[25] This phenomenon was explained by a hypothesis assuming that the therapeutic action is not a result of the action of drugs as such, but the organism's response to their permanent presence in the body, leading to adaptive changes. This hypothesis based on the stud-

---

[23] B. Haenisch, H. Bönisch, *Depression and Antidepressants: Insights from Knockout of Dopamine, Serotonin or Noradrenaline Re-uptake Transporters*, "Pharmacology & Therapeutics" 2011, vol. 129, pp. 352–368.

[24] J.J. Schildkraut, *The Catecholamine Hypothesis of Affective Disorders: A Review of Supporting Evidence*, "American Journal of Psychiatry" 1965, vol. 122, pp. 509–522.

[25] I. Oswald, V. Brezinova, D.L. Dunleavy, *On the Slowness of Action of Tricyclic Antidepressant Drugs*, "British Journal of Psychiatry" 1972, vol. 120, pp. 673–677.

ies of Vetulani and Sulser, demonstrating that the chronic administration of various antidepressant drugs of various types and the chronic application of electroconvulsive treatments cause a depression of the responsiveness of the cyclic AMP generating system in specific brain regions, but not earlier than after two weeks of treatment.[26, 27, 28] This hypothesis, named beta-downregulation hypothesis, dominated the scene of antidepressant research for a few years, until it was found that the phenomenon of decline in beta-adrenoceptor density does not appear after application of several effective antidepressants, particularly SSRI,[29] and that chronic antidepressant treatments induce rather activation than inhibition of intracellular signaling cascade.[30]

Recently, the disturbances in monoaminergic transmission in depression have been treated rather as epiphenomena, and not causative factors. Depression and responses to antidepressant therapy seem to be controlled by as yet not fully recognized final physiological pathways whose functions are modulated by monoamines. The activity of particular monoaminergic tracts seem to be a primary phenomenon that modulates the secondary, causatively important events, such as changes in gene expression and protein synthesis, eventually leading to depression and modulating the responses to antidepressants.[31]

---

[26] J. Vetulani, F. Sulser, *Action of Various Antidepressant Treatments Reduces Reactivity of Noradrenergic Cyclic AMP-generating System in Limbic Forebrain*, "Nature" 1975, vol. 257, pp. 495–496.

[27] J. Vetulani, R.J. Stawarz, J.V. Dingell, F. Sulser, *A Possible Common Mechanism of Action of Antidepressant Treatments: Reduction in the Sensitivity of the Noradrenergic Cyclic AMP Gererating System in the Rat Limbic Forebrain*, "Naunyn-Schmiedeberg's Archives of Pharmacology" 1976, vol. 293, pp. 109–114.

[28] J. Vetulani, R.J. Stawarz, F. Sulser, *Adaptive Mechanisms of the Noradrenergic Cyclic AMP Generating System in the Limbic Forebrain of the Rat: Adaptation to Persistent Changes in the Availability of Norepinephrine (NE)*, "Journal of Neurochemistry" 1976, vol. 27, pp. 661–666.

[29] I. Nalepa, J. Vetulani, *Enhancement of the Responsiveness of Cortical Adrenergic Receptors by Chronic Administration of the 5-hydroxytryptamine Uptake Inhibitor Citalopram*, "Journal of Neurochemistry" 1993, vol. 60, pp. 2029–2035.

[30] D.B. Menkes, M.M. Rasenick, M.A. Wheeler, M.W. Bitensky, *Guanosine Triphosphate Activation of Brain Adenylate Cyclase: Enhancement by Long-term Antidepressant Treatment*, "Science" 1983, vol. 219, pp. 65–67.

[31] H.K. Manji, W.C. Drevets, D.S. Charney, *The Cellular Neurobiology of Depression*, "Natural Medicine" 2001, vol. 7, pp. 541–547.

The antidepressants currently in use in clinics almost without exception affect the monoaminergic transmission. Our knowledge of which type of monoaminergic transmission is most effectively facilitated by a given drug permits us to choose a proper therapy depending on the type of depression and its symptoms.

1.1 Other Neurotransmitters

Several neurotransmitter systems, beyond the monoaminergic, also display abnormalities in depression, and it has been postulated that those disturbances also might be of etiological importance.

1.2 Acetylcholine

The cholinergic system is one of those that are also implicated in depression.[32] Its involvement was postulated initially by Janowsky el al. in 1972,[33] but their hypothesis did not gather convincing proofs at that time. Nevertheless, the cholinergic system plays a significant role in the control of several cerebral functions, such as alertness, attention, cognition and memory, which are impaired in depression. Particularly, cholinergic deficits may be responsible for depression--associated cognitive deficits, and such deficits may also be responsible for impairment of hippocampal neurogenesis,[34, 35] regarded as one of essential consequence or cause of depression. It seems that the

---

[32] G. Dagytė, J.A. Den Boer, A. Trentani, *The Cholinergic System and Depression*, "Behavioural Brain Research" 2011, vol. 221, pp. 574–582.

[33] D.S. Janowsky, M.K. el-Yousef, J.M. Davis, H.J. Sekerke, *A Cholinergic-adrenergic Hypothesis of Mania and Depression*, "Lancet" 1972, vol. 2, pp. 632–635.

[34] C.M. Cooper-Kuhn, J. Winkler, H.G. Kuh, *Decreased Neurogenesis After Cholinergic Forebrain Lesionin the Adult Rat*, "Journal of Neuroscience Research" 2004, vol. 77, pp. 155–165.

[35] K. van der Borght, J. Mulder, J.N. Keijser, B.J. Eggen, P.G. Luiten, E.A. van der Zee, *Input from the Medial Septum Regulates Adult Hippocampal Neurogenesis*, "Brain Research Bulletin" 2005, vol. 67, pp. 117–125.

mild facilitation of cholinergic transmission, by, e.g., administration of choline precursors of lecithin type, may be considered in depressed patients with exacerbated, annoying cognitive dysfunctions.

1.3 Glutamate

Glutamate is the principal excitatory neurotransmitter in the mammalian brain, and its involvement in psychiatric disorder has been investigated for several years.[36] Glutamatergic neurons represent approximately one third of all forebrain neurons, and glutamatergic receptors are present in high density in the cortex, hippocampus, caudate nucleus, thalamic nuclei and cerebellum.[37] Preclinical studies on antidepressant potential of glutamatergic neurons were carried out in the last century, and their major achievement was the discovery of the rapid clinical action of ketamine.[38] However, because of unknown reasons, this line of research ceased to arouse interest among researchers for some time, and its revival has only taken place in recent years.

The glutamatergic system is an interesting target in the search for new antidepressants as its activity may be modulated at many sites. The system operates through a variety of receptors, both ionotropic (AMPA, NMDA and others) and metabotropic, that play various functions and are modulated by various ligands. A glutamatergic synapse is typically formed by three cells: an astrocyte, a presynaptic neuron and a postsynaptic neuron.[39] Glutamate formed in the

---

[36] D.C. Javitt, S.R. Zukin, *The Role of Excitatory Amino Acids in Neuropsychiatric Illness*, "The Journal of Neuropsychiatry & Clinical Neuroscience" 1990, vol. 2, pp. 44–52.
[37] G. Rajkowska, *Histopathology of the Prefrontal Cortex in Major Depression: What Does It Tell Us About Dysfunctional Monoaminergic Circuits?*, "Progress in Brain Research" 2000, vol. 126, pp. 397–412.
[38] R.M. Berman, A. Cappiello, A. Anand, D.A. Oren, G.R. Heninger, D.S. Charney, J.H. Krystal, *Antidepressant Effects of Ketamine in Depressed Patients*, "Biological Psychiatry" 2000, vol. 47, pp.351–354.
[39] C.A. Zarate, J. Quiroz, J. Payne, H.K. Manji, *Modulators of the Glutamatergic System: Implications for the Development of Improved Therapeutics in Mood Disorders*, "Psychopharmacology Bulletin" 2002, vol. 36, pp. 35–83.

presynaptic neuron is packed into the presynaptic vesicles by vesicular transporters, released by exocytosis to the synaptic cleft, and taken up by specific transporters into the neurons and the astrocyte.[40] Each of these processes may be modified by drugs. Thus, the preliminary findings suggest that the decreased expression of the vesicular glutamate transporter VGLUT1 is associates with appearance of "depressive-like" type of behaviour,[41] and antidepressants such as paroxetine, fluoxetine and desipramine enhance the expression of this transporter in the structures involved in processing of emotions.[42] Modification of glutamate release may affect mood: preliminary studies indicate that riluzol, the glutamate release inhibitor employed in ALS therapy, is effective in treating depression.[43, 44]

The participation of particular types of glutamate receptors in potential antidepressant action (at least 23 of receptor subtypes were investigated in this direction) is brilliantly summed up in the review of Machado-Viera et al.[45] The studies on the participation of glutamatergic system in the control of mood and on possibilities of its modulations as potential antidepressant therapies appear very promising.

---

[40] See R. Machado-Vieira, G. Salvadore, L.A. Ibrahim, N. Diaz-Granados, C.A. Zarate Jr., *Targeting Glutamatergic Signaling for the Development of Novel Therapeutics for Mood Disorders*, "Current Pharmaceutical Design" 2009, vol. 15, pp. 1595–1611.

[41] R.M. Tordera, S. Totterdell, S.M. Wojcik, N. Brose, N. Elizalde, B. Lasheras, J. Del Rio, *Enhanced Anxiety, Depressive-like Behaviour and Impaired Recognition Memory in Mice with Reduced Expression of the Vesicular Glutamate Transporter 1 (VGLUT1)*, "European Journal of Neuroscience" 2007, vol. 25, pp. 281–290.

[42] R.M. Tordera, Q. Pei, T. Sharp, *Evidence for Increased Expression of the Vesicular Glutamate Transporter, VGLUT1, by a Course of Antidepressant Treatment*, "Journal of Neurochemistry" 2005, vol. 94, pp. 875–883.

[43] C.A. Zarate Jr., J.L. Payne, J. Quiroz, J. Sporn, K.K. Denicoff, D. Luckenbaugh, D.S. Charney, H.K. Manji, *An Open-label Trial of Riluzole in Patients with Treatment-resistant Major Depression*, "American Journal of Psychiatry" 2004, vol. 161, pp. 171–174.

[44] G. Sanacora, S.F. Kendell, Y. Levin, A.A. Simen, L.R. Fenton, V. Coric, J.H. Krystal, *Preliminary Evidence of Riluzole Efficacy in Antidepressant-treated Patients with Residual Depressive Symptoms*, "Biological Psychiatry" 2007, vol. 61, pp. 822–825.

[45] R. Machado-Vieira, G. Salvadore, L.A. Ibrahim, N. Diaz-Granados, C.A. Zarate Jr., *Targeting..., op. cit.*

1.4 Tachykinins – Substance P

Substance P (SP) is the first neuropeptide discovered and is an important pain regulator. New findings suggest that it may mediate not only physical pain, nociception, but also mental pain – depression. SP belongs, together with neurokinin A and neurokinin B, to tachykinins, neuropeptides present almost exclusively in neurons and playing a role of neurotransmitters or neuromodulators in the mammalian central nervous system.[46]

SP and other tachykinins were found in several forebrain, midbrain and brain stem areas, particularly in structures modulating stress, anxiety and mood, such as cingulate cortex, striatum, nucleus accumbens, septum, amygdala, various hypothalamic areas, periaqueductal gray, raphe nuclei, locus coeruleus, various parabrachial nuclei and the nucleus of the solitary tract.[47, 48, 49]

Several, though not all, human studies demonstrated an elevated SP level in the cerebrospinal fluid, serum and sweat of patients with major depression. The attempts with the administration of tachykinin receptor antagonists as antidepressants were carried out for some time, but eventually the negative result of large multicenter trial of phase III[50] caused the withdrawal of several pharmaceutical companies, firstly Merck, followed by Pfizer, Takeda-Abbotti and Roche, from this research, leaving only Glaxo on the battlefield.

---

46 M. Otsuka, K. Yoshioka, *Neurotransmitter Functions of Mammalian Tachykinins*, "Physiological Reviews" 1993, vol. 73, pp. 229–308.

47 T. Hökfelt, E. Kuteeva, D. Stanic, A. Ljungdahl, *The Histochemistry of Tachykinin Systems in the Brain*, [in:] *Handbook of Experimental Pharmacology*, ed. P. Holzer, Springer-Verlag, Berlin–Heidelberg 2004, pp. 63–120.

48 Y. Nakaya, T. Kaneko, R. Shigemoto, S. Nakanishi, N. Mizuno, *Immunohistochemical Localization of Substance P Receptor in the Central Nervous System of the Adult Rat*, "The Journal of Comparative Neurology" 1994, vol. 347, pp. 249–274.

49 C.W. Shults, R. Quirion, B. Chronwall, T.N. Chase, T.L. O'Donohue, *A Comparison of the Anatomical Distribution of Substance P and Substance P Receptors in the Rat Central Nervous System*, "Peptides" 1984, vol. 5, pp. 1097–1128.

50 M. Keller, S. Montgomery, W. Ball, M. Morrison, D. Snavely, G. Liu, R. Hargreaves, J. Hietala, C. Lines, K. Beebe, S. Reines, *Lack of Efficacy of the Substance p (Neurokinin1 Receptor) Antagonist Aprepitant in the Treatment of Major Depressive Disorder*, "Biological Psychiatry" 2006, vol. 59, pp. 216–223.

Nevertheless several studies, described in the review of Ebner et al.,[51] suggest that tachykinins may be involved in etiology of depression, and SP antagonists form a potential source of antidepressant drugs. As the most popular antidepressants still display efficiency lower than expected by patients and doctors,[52] the attempts are undertaken to increase their efficiency by combining them with drugs with different action mechanism. Hence the trials of a combination of SSRI's with tachykinin antagonists.[53, 54] Unfortunately, the expectations of obtaining clinically more efficient antidepressants in this way have not as yet been fulfilled.

1.5 Stress

Mood changes are often precipitated or aggravated by stressful conditions,[55] and therefore stress is regarded as one of the causes of depression. Responses to stress are associated with the release of a neuropeptide, corticotropin-releasing hormone (CRH), also known as corticoliberin or CRF, that stimulates the pituitary synthesis of ACTH. and – consequently – enhances the production and release of adrenal corticosteroids. The receptors for corticosteroids are present at high densities in the hippocampus, the structure regarded as critically important in the etiology of depression. The excessive stimula-

[51] K. Ebner, S.B. Sartori, N. Singewald, *Tachykinin Receptors as Therapeutic Targets in Stress-related Disorders*, "Current Pharmaceutical Design" 2009, vol. 15, pp. 1647–1674.

[52] M.E. Thase, A.J. Rush, *When at First You Don't Succeed: Sequential Strategies for Antidepressant Nonresponders*, "The Journal of Clinical Psychiatry" 1997, vol. 58 Suppl 13, pp. 23–29.

[53] T. Ryckmans, L. Balançon, O. Berton, C. Genicot, Y. Lamberty, B. Lallemand, P. Pasau, N. Pirlot, L. Quéré, P. Talaga, *First dual NK(1) Antagonists-serotonin Reuptake Inhibitors: Synthesis and SAR of a New Class of Potential Antidepressants*, "Bioorganic & Medicinal Chemistry Letters" 2002, vol. 12, pp. 261–264.

[54] T. Ryckmans, O. Berton, R. Grimée, T. Kogej, Y. Lamberty, P. Pasau, P. Talaga, C. Genicot, *Dual NK(1) Antagonists – Serotonin Reuptake Inhibitors as Potential Antidepressants. Part 2: SAR and Activity of Benzyloxyphenethyl Piperazine Derivatives*, "Bioorganic & Medicinal Chemistry Letters" 2002, vol. 12, pp. 3195–3198.

[55] P.W. Gold, G.P. Chrousos, *Organization of the Stress System and Its Dysregulation in Melancholic and Atypical Depression: High vs Low CRH/NE States*, "Molecular Psychiatry" 2002, vol. 7, pp. 254–275.

tion of corticoid receptors in the hippocampus results in the damaging of hippocampal neurons.[56]

CRH and related peptides, urocortin 1 (UCN1), 2 (UCN 2), and 3 (UCN3], are important mediators of central and peripheral stress responses.[57] In the central nervous system, these peptides participate in the regulation of behavioural, cognitive, autonomic and immunologic responses to aversive stimuli.[58, 59]

Abnormalities in neurotransmission mediated by CRH and in signal transduction involving CRH1 receptors were regarded as critical mechanisms of stress pathophysiology leading to depression. The markers of hyperactivity of CRH neurotransmission are high levels of CRH in the cerebrospinal fluid and functional disturbances of the hypothalamic-pituitary-adrenal axis (HPA).[60]

After the ground-breaking discovery of Vale et al.,[61] who isolated and sequenced CRH, further studies supported the idea that CRH played the primary role in the pathophysiology of depression. Following cloning of CRH1 and CRH2 receptors, several pharmaceutical companies began to synthesize the antagonists of those receptors as potential antidepressant and anxiolytic agents.[62] Until

---

[56] B. McEwen, *Stress and Hippocampal Plasticity*, "Current Opinion in Neurobiology" 1999, vol. 5, pp. 205–216.

[57] R.L. Hauger, V. Risbrough, R.H. Oakley, J.A. Olivares-Reyes, F.M. Dautzenberg, *Role of CRF Receptor Signaling in Stress Vulnerability, Anxiety, and Depression*, "Annals of the New York Academy of Science" 2009, vol. 1179, pp. 120–143.

[58] R.L. Hauger, D.E. Grigoriadis, M.F. Dallman, P.M. Plotsky, W.W. Vale, F.M. Dautzenberg, *International Union of Pharmacology. XXXVI. Current Status of the Nomenclature for Receptors for Corticotropin-releasing Factor and Their Ligands*, "Pharmacological Reviews" 2003, vol. 55, pp. 21–26.

[59] T.L. Bale, W.W. Vale, *CRF and CRF Receptors: Role in Stress Responsivity and Other Behaviours*, "Annual Review of Pharmacology and Toxicology" 2004, vol. 44, pp. 525–557.

[60] F. Holsboer, M. Ising, *Central CRH System in Depression and Anxiety – Evidence from Clinical Studies with CRH1 Receptor Antagonists*, "European Journal of Pharmacology" 2008, vol. 583, pp. 350–357.

[61] W. Vale, J. Spiess, C. Rivier, J. Rivier, *Characterization of a 41-residue Ovine Hypothalamic Peptide that Stimulates Secretion of Corticotropin and Beta-endorphin*, "Science" 1981, vol. 213, pp. 1394–1397.

[62] D.E. Grigoriadis, *The Corticotropin-releasing Factor Receptor: A Novel Target for the Treatment of Depression and Anxiety-related Disorders*, "Expert Opinion on Therapeutic Targets" 2005, vol. 9, pp. 651–684.

now, however, no successful antidepressant from this group has been found.

Preclinical studies demonstrated that stress may induce structural changes in the hippocampus,[63, 64, 65] and that those changes may be attenuated by a successful antidepressant treatment.[66, 67, 68]

1.6 Neonatal Stress

Recently attention has been focused on the effect of stress effects in the early period of life. Epidemiological studies have revealed that exposure to stress in the prenatal period and in the first stages of postnatal life is associated with the liability to depression in adult life. The early negative consequences of neonatal stress are deficits, both cognitive (in memory, cognition, executive functions) and emotional (in processing of rewarding and social stimuli, emotional control). The analysis of data suggests that the most vulnerable to neonatal stress are complex cognitive functions of a higher order, associated with the brain areas that develop slowly in the postnatal period. The area particularly susceptible is the amygdala, and several deficits, espe-

[63] J.E. Malberg, R.S. Duman, *Cell Proliferation in Adult Hippocampus is Decreased by Inescapable Stress: Reversal by Fluoxetine Treatment*, "Neuropsychopharmacology" 2003, vol. 28, pp. 1562–1571.

[64] K. Pham, J. Nacher, P.R. Hof, B.S. McEwen, *Repeated Restraint Stress Suppresses Neurogenesis and Induces Biphasic PSA-NCAM Expression in the Adult Rat Dentate Gyrus*, "European Journal of Neuroscience" 2003, vol. 17, pp. 879–886.

[65] E. Vermetten, J.D. Bremner, *Circuits and Systems in Stress. I. Preclinical Studies*, "Depress Anxiety" 2002, vol. 15, pp. 126–147.

[66] R.S. Duman, S. Nakagawa, J. Malberg, *Regulation of Adult Neurogenesis by Antidepressant Treatment*, "Neuropsychopharmacology" 2001, vol. 25, pp. 836–844.

[67] T.D. Perera, J.D. Coplan, S.H. Lisanby, C.M. Lipira, M. Arif, C. Carpio, G. Spitzer, L. Santarelli, B. Scharf, B. Hen R, G. Rosoklija, H.A. Sackeim, A.J. Dwork, *Antidepressant-induced Neurogenesis in the Hippocampus of Adult Nonhuman Primates*, "Journal of Neuroscience" 2007, vol. 27, pp. 4894–4901.

[68] L. Santarelli, M. Saxe, C. Gross, A. Surget, F. Battaglia, S. Dulawa, N. Weisstaub, J. Lee, R. Duman, O. Arancio, C. Belzung, R. Hen, *Requirement of Hippocampal Neurogenesis for the Behavioral Effects of Antidepressants*, "Science" 2003, vol. 301, pp. 805–809.

cially in the affective domain, may persist long after disappearance of stressful stimuli and increase the risk of later psychopathological changes.[69] The perinatal stress has been employed to construct an animal model of depression.[70]

Although early stress is regarded as a detrimental condition, the hypothesis has appeared that it may be biologically beneficial: the stress in the earliest periods of life may play an important adaptive role, preparing the individual to survive in a hostile environment. Only if the later periods are more friendly do the adaptive mechanisms formed earlier become counteradaptive and pathological. According to this "mismatch hypothesis" it is not neonatal stress itself, but the unmatching of the early formed program to the later, better than expected life conditions, that is the source of depression.[71]

1.7 Stress and BDNF

The brain-derived neurotrophic factor (BDNF) is thought to be the most potent antistressor agent, as it counteracts the stress-induced neuronal damage. Stress causes damage or atrophy of neurons in several brain areas, particularly in the hippocampus, containing dense population of receptors of glucocorticoids, the main stress hormones of adrenals.[72, 73, 74] The induction of anatomical changes by stress gave

---

[69] P. Pechtel, D.A. Pizzagalli, *Effects of Early Life Stress on Cognitive and Affective Function: an Integrated Review of Human Literature*, "Psychopharmacology (Berl)" 2011, vol. 214, pp. 55–70.

[70] M.V. Schmidt, X.D. Wang, O.C. Meijer, *Early Life Stress Paradigms in Rodents: Potential Animal Models of Depression?*, "Psychopharmacology (Berl)" 2011, vol. 214, pp. 131–140.

[71] M.V. Schmidt, *Animal Models for Depression and the Mismatch Hypothesis of Disease*, "Psychoneuroendocrinology" 2011, vol. 36, pp. 330–338.

[72] R. Sapolsky, *Glucocorticoids and Atrophy of the Human Hippocampus*, "Science" 1996, vol. 273, pp. 749–750.

[73] R. Sapolsky, *Depression, Antidepressants, and the Shrinking Hippocampus*, "Proceedings of the National Academy of Science" 2001, vol. 98, pp. 12320–12322.

[74] R. Duman, *Role of Neurotrophic Factors in the Etiology and Treatment of Mood Disorders*, "NeuroMolecular Medicine" 2004, vol. 5, pp. 11–25.

an assumption to investigate the role of trophic factors, and the results showed that stresses of various kinds depress the expression of BDNF in the hippocampus.[75]

BDNF expression is blocked by corticosterone administration, but this hormone is not the only factor controlling BDNF expression. Although adrenalectomy causes elevation of the BDNF concentration, it does not antagonize completely all effects of stress.[76] The effect of stress on BDNF is partially prevented by inhibition of 5HT2A receptors that control GABA interneurons in the hippocampus.[77] There is also evidence of the inhibition of hippocampal expression of BDNF by interleukin-1.[78] Stress also suppresses the levels of other neurotrophins, NT-3 and NGF,[79] and of the vascular endothelium growth factor (VEGF) that was found to promote neuronal proliferation in the adult hippocampus.[80]

Antidepressant therapies increase the expression of BDNF,[81] whose level is lowered in depressed patients. BDNF expression may also be enhanced by physical activity, but usually it is very difficult to persuade patients with major depression to exercise.

---

[75] R.S. Duman, L.M. Monteggia, *A Neurotrophic Model for Stress-related Mood Disorders*, "Biological Psychiatry" 2006, vol. 59, pp. 1116–1127.

[76] M.A. Smith, S. Makino, R. Kvetnansky, R.M. Post, *Stress and Glucocorticoids Affect the Expression of Brain-derived Neurotrophic Factor and Neurotrophin-3 mRNAs in the hippocampus*, "Journal of Neuroscience" 1995, vol. 15, pp. 1768–1777.

[77] V.A. Vaidya, G.J. Marek, G.K. Aghajanian, R.S. Duman, *5-HT2A Receptor-mediated Regulation of Brain-derived Neurotrophic Factor mRNA in the Hippocampus and the Neocortex*, "Journal of Neuroscience" 1997, vol. 17, pp. 2785–2795.

[78] R.M. Barrientos, D.B. Sprunger, S. Campeau, E.A. Higgins, L.R. Watkins, J.W. Rudy, S.F. Maier, *Brain-derived Neurotrophic Factor mRNA Downregulation Produced by Social Isolation is Blocked by Intrahippocampal Interleukin-1 Receptor Antagonist*, "Neuroscience" 2003, vol. 121, pp. 847–853.

[79] T. Ueyama, Y. Kawai, K. Nemoto, M. Sekimoto, S. Toné, E. Senba, *Immobilization Stress Reduced the Expression of Neurotrophins and Their Receptors in the Rat Brain*, "Neuroscience Research" 1997, vol. 28, pp. 103–110.

[80] T. Palmer, A.R. Willhoite, F.H. Gage, *Vascular Niche for Adult Hippocampal Neurogenesis*, "The Journal of Comparative Neurology" 2000, vol. 425, pp. 479–494.

[81] M. Nibuya, S. Morinobu, R.S. Duman, *Regulation of BDNF and trkB mRNA in Rat Brain by Chronic Electroconvulsive Seizure and Antidepressant Drug Treatments*, "Journal of Neuroscience" 1995, vol. 15, pp. 7539–7547.

It is worth mentioning that BDNF appears in two polymorphic forms, with valine or methionine at position 55. The methionine form is abnormally packed in the neuron and less easily released, and therefore less effective,[82] so individuals carrying out Val66Met polymorphism have a higher risk of development of some psychiatric conditions.

1.8 Neuroplasticity

Neuroplasticity is defined as the ability of the brain to reorganize its circuits and structures in response to stimuli. This is an essential property of the brain, permitting it to play its basic role – to function as the organ for survival.[83] Neuroplasticity is responsible for the shrinking of the dendritic tree, its atrophy and, eventually, for the apoptosis of neurons that are not used, and for development of neurons challenged to respond by environmental conditions. One of the most important neuroplastic processes is learning, and the most positive neuroplastic response is the formation of new neurons – neurogenesis, which in humans was thought to appear only in the hippocampus, but recently has also been found in the olfactory bulb. It is postulated that depression in humans is associated with the acceleration of neurodegeneration and the reduction of neurogenesis in the hippocampus.[84] Such conditions are induced by stress, and counteracted by the activity of BDNF. The concept of depression as the loss of neuroplasticity in the hippocampus turned the

---

[82] M.F. Egan, M. Kojima, J.H. Callicott, T.E. Goldberg, B.S. Kolachana, A. Bertolino, E. Zaitsev, B. Gold, D. Goldman, M. Dean, B. Lu, D.R. Weinberger, *The BDNF Val-66met Polymorphism Affects Activity-dependent Secretion of BDNF and Human Memory and Hippocampal Function*, "Cell" 2003, vol. 112, pp. 257–269.

[83] E.O. Wilcon, Consilience: *The Unity of Knowledge*, Alfred A. Knopf, New York 1998.

[84] M. Maes, *Inflammatory and Oxidative and Nitrosative Stress Pathways Underpinning Chronic Fatigue, Somatization and Psychosomatic Symptoms*, "Current Opinion in Psychiatry" 2009, vol. 22, pp. 75–83.

attention to the fact that stimulation of neuroplasticity may be an important component in the action of antidepressants. This was found true for most antidepressant drugs, for some of them, e.g. tianeptine, enhancement of neuroplasticity may be the main mechanism of antidepressant action, exerted without interference with neurotransmitter systems.

Untreated depression leads to negative neuroplastic effects: the atrophy and extinction of neurons. There is a nonpharmacological possibility of increasing patients' neuroplasticity by exposing them to an enriched environment, and it seems plausible that contact with a psychotherapist may, besides all other, play a function of enriched environment.

1.9 The Exaggerated Repair Response Hypothesis of Depression

One of the last, very interesting concepts of depression was presented by Wagner-Smith and Markou, who postulated that depression results from a not fully successful correction response to stress-induced brain microinjuries.[85] Their theoretical model assumes that stressful life events induce small brain injuries, associated with an elevation of cortisol level. The chronic exposure to glucocorticoids causes the shortening and decrease of arborization of apical dendrites of hippocampal neurons, similar to those observed after chronic exposure to stress.[86] Those impairments should, in turn, induce repair response, starting from inflammatory reactions (neuroinflammation), aimed at the removal of injured cellular fragments, and followed by the initiation of spontaneous regenerative responses, involving neurotrophins, pos-

[85] K. Wager-Smith, A. Markou, *Depression: a Repair Response to Stress-induced Neuronal Microdamage that Can Grade into a Chronic Neuroinflammatory Condition?*, "Neuroscience & Biobehavioral Reviews" 2011, vol. 35, pp. 742–764.

[86] C.D. Conrad, K.J. McLaughlin, J.S. Harman, C. Foltz, L. Wieczorek, E. Lightner, R.L. Wright, *Chronic Glucocorticoids Increase Hippocampal Vulnerability to Neurotoxicity Under Conditions that Produce CA3 Dendritic Retraction But Fail to Impair Spatial Recognition Memory*, "Journal of Neuroscience" 2007, vol. 27, pp. 8278–8285.

sibly accompanied by neurogenesis. The neuroinflammatory transmitters released during the process of repair may cause pathological behaviour and – as the authors propose – cause mental pain analogous to physical pain encountered during wound healing. Although the comparison of mental and physical pain seems risky, it might be mentioned that opioids, strong analgesic compounds and euphoriants, were used in the treatment of depression before introduction of modern antidepressants,[87] and even recently have been suggested as effective in depression.[88] Similarly, rapid antidepressant action is exerted by a strong dissociative analgesic, ketamine.[89]

The idea that depression is associated with brain impairment is corroborated by findings that in humans it often appears after traumatic brain injury,[90] even relatively mild,[91] and after a stroke.[92] It may also be speculated that senile depression results from the accumulation of silent microinfarcts, appearing as white matter intensities in fNMR images.[93]

If the healing process is successful, depression remits, but frequently appearing acute phases involving inflammatory processes accompanied by mental pain (i.e. mood lowering) may turn into a pathological chronic state. Although, in some points, the above hypothesis uses scientifically unfounded analogies between physical and mental

[87] T.A. Ban, *Pharmacotherapy of Depression: a Historical Analysis*, "Journal of Neural Transmission" 2001, vol. 108, pp. 707–716.

[88] P.L. Tenore, *Psychotherapeutic Benefits of Opioid Agonist Therapy*, "Journal of Addictive Diseases" 2008, vol. 27, pp. 49–65.

[89] P. Skolnick, P. Popik, R. Trullas, *Glutamate-based Antidepressants: 20 Years On*, "Trends in Pharmacological Sciences" 2009, vol. 30, pp. 563–569.

[90] C.H. Bombardier, J.R. Fann, N.R. Temkin, P.C. Esselman, J. Barber, S.S. Dikmen, *Rates of Major Depressive Disorder and Clinical Outcomes Following Traumatic Brain Injury*, "Journal of American Medical Association" 2010, vol. 303, pp. 1938–1945.

[91] L.M. Ryan, D.L. Warden, *Post Concussion Syndrome*, "International Review of Psychiatry" 2003, vol. 15, pp. 310–316.

[92] G.L. Lenzi, M. Altieri, I. Maestrini, *Post-stroke Depression*, "Revue Neurologique Société de Neurologie de Paris." 2008, vol. 164, pp. 837–840.

[93] M. Santos, E. Kövari, P.R. Hof, G. Gold, C. Bouras, P. Giannakopoulos, *The Impact of Vascular Burden on Late-life Depression*, "Brain Research Reviews" 2009, vol. 62, pp. 19–32.

pain (but it might be mentioned that substance P and its receptors are supposed to be engaged in both phenomena[94]), it explains why treatments and drugs that increase neuroplasticity exert beneficial effects.

An interesting idea of the authors of the concept of depression as the repair response to stress is that the very high vulnerability of the brain to stress observed (limited mainly to the hippocampus, prefrontal cortex, and amygdala) serves to eliminate the circuits that inadequately react to stress and permits adaptive reconstruction of the brain.

2. Structural Changes

The introduction of neuroimaging permitted the investigation of structural changes occurring in the course of psychiatric diseases. These changes take place in specific brain areas. In patients with major depression the changes in volume were regularly observed in the prefrontal cortex, cingular cortex and the temporal lobe.

2.1 Frontal Cortex

A decrease in the volume of frontal cortex due to a reduction in the cortical thickness, neuronal sizes, and neuronal and glial densities was observed in several sites.[95] The most prominent observed changes were the decreases of grey matter in the left anterior cingulate cortex located ventrally to the genu of corpus callosum (so-called subgenual cingular cortex, Brodmann area 25, cg25). Those changes, reaching up to 48%, were described for the first time by Drevets.[96] It appears

---

[94] C.L. DeVane, *Substance P: A New Era, a New Role*, "Pharmacotherapy" 2001, vol. 21, pp. 1061–1069.

[95] G. Rajkowska, J.J. Miguel-Hidalgo, J. Wei, G. Dilley, S.D. Pittman, H.Y. Meltzer, J.C. Overholser, B.L. Roth, C.A. Stockmeier, *Morphometric Evidence for Neuronal and Glial Prefrontal Cell Pathology in Major Depression*, "Biological Psychiatry" 1999, vol. 45, pp. 1085–1098.

[96] W.C. Drevets, J.L. Price, J.R. Simpson Jr, R.D. Todd, T. Reich, M. Vannier, M.E. Raichle, *Subgenual Prefrontal Cortex Abnormalities in Mood Disorders*, "Nature" 1997, vol. 386, pp. 824–827.

that they develop early in the course of disease, and appear also in young healthy persons with high familial risk of depression.[97] Interestingly, contrasting to the dramatic decrease in volume, this very region displays the highest neuronal activity in depression. It is possible that the changes in volume are related to the hyperactivity of one of the elements of anatomical limbic-cortical-striatal-pallidal-thalamic (LCSPT) circuit (vide infra), leading to the activation of the whole system and resulting in the extinction of neurons due to excitotoxity, particularly as the areas in question have a high density of glucocorticoid receptors.

In depressed patients, the grey mater volume is also reduced in frontoorbital and ventromedial prefrontal cortex.[98]

2.2 Hippocampus

Several, but not all, studies have demonstrated evident decreases in the hippocampal volume in depression.[99] The changes are most easily observed in older patients with recurrent, treatment-resistant depression. A meta-analysis of over 2000 cases has confirmed that the volume of the hippocampus of depressed patients is reduced, but only in those in which the disease lasted for at least two years and who experienced more than one depression episode. The volume reduction was observed in patients of all age groups. It appears that, in contrast to changes in subgenual cingulate, the reduction of the hippocampal volume commences after appearance of the disease, and thus is its

---

97 K.N. Botteron, M.E. Raichle, W.C. Drevets, A.C. Heath, R.D. Todd, *Volumetric Reduction in Left Subgenual Prefrontal Cortex in Early Onset Depression*, "Biological Psychiatry" 2002, vol. 51, pp. 342–344.

98 W.C. Drevets, J.L. Price, *Neuroimaging and Neuropathological Studies of Mood Disorders*, [in:] *Biology of Depression: From Novel Insights to Therapeutic Strategies*, ed. J.W.M. Licinio, Wiley-VCH Verlag GmbH & Co., Weinheim 2005.

99 S. Campbell, M. Marriott, C. Nahmias, G.M. MacQueen, *Lower Hippocampal Volume in Patients Suffering from Depression: a Meta-analysis*, "American Journal of Psychiatry" 2004, vol. 161, pp. 598–607.

result, and not the cause.[100] An interesting post-mortem finding is the significant increase in the packing density of glia, pyramidal neurons, and granule cell neurons, and thus the reduction of neuropil in the hippocampi of depressed patients.[101]

The studies repeated on the same patients with major depression twice, at the interval of three years, shed new light on the progress of disease. They demonstrated that in the course of depression there is a decline in the density of gray matter in the hippocampus, anterior cingulate, left amygdala and right dorsomedial prefrontal cortex. Those changes were greater in the patients that had no remission in the period between tests. This result confirms the conclusions drawn from animal experiments that in depression-associated brain areas there appear stress-induced negative neuroplastic changes.[102]

The studies on changes in the volume of amygdala and basal ganglia in depression did not yield unequivocal results.[103]

3. Depression as a Result of the Malfunction of Neuronal Circuits

The modern view of the etiology of depression abandons the attention on the impairment of the transmission of signals between a pair of neurons through a synapse, but assumes that depression results

[100] M.C. McKinnon, K. Yucel, A. Nazarov, G.M. MacQueen, *A Meta-analysis Examining Clinical Predictors of Hippocampal Volume in Patients with Major Depressive Disorder*, "Journal of Psychiatry and Neuroscience" 2009, vol. 34, pp. 41–54.

[101] C.A. Stockmeier, G.J. Mahajan, L.C. Konick, J.C. Overholser, G.J. Jurjus, H.Y. Meltzer, H.B. Uylings, L. Friedman, G. Rajkowska, *Cellular Changes in the Postmortem Hippocampus in Major Depression*, "Biological Psychiatry" 2004, vol. 56, pp. 640–650.

[102] T.S. Frodl, N. Koutsouleris, R. Bottlender, C. Born, M. Jäger, I. Scupin, M. Reiser, H.J. Möller, E.M. Meisenzahl, *Depression-related Variation in Brain Morphology over 3 Years: Effects of Stress?*, "Archives of General Psychiatry" 2008, vol. 65, pp. 1156–1165.

[103] W.C. Drevets, J.L. Price, J.R. Simpson Jr., R.D. Todd, T. Reich, M. Vannier, M.E. Raichle, *Subgenual Prefrontal Cortex Abnormalities in Mood Disorders*, "Nature" 1997, vol. 386, pp. 824–827.

from the breakdown in circuits. This concerns the brain circuit involved in the evaluative, expressive and experiential aspects of emotional behaviour.[104] This is the so-called limbic-cortical-striatal-pallidal-thalamic circuits (LCSPT), formed by connections between the orbital and medial prefrontal cortex, amygdala, subiculum, ventromedial striatum, ventral pallidum and thalamic nuclei (mediodorsal and midline).[105] Disturbances that in various ways affect the transmission in this circuit cause pathological emotional symptoms that fulfil the criteria of major depression.[106]

This approach paves the way for the nonpharmacological biological treatment of depression, by direct manipulation of this circuit. The vulnerable points in this circuit may be presently detected with neuroimaging techniques. A good example is the finding of the hyperactivity of cg25 area in depression. Such hyperactive areas have become a natural target for neurosurgery. The recent techniques of deep brain stimulation, though invasive, are reversible, and attempts at their application have become more and more common. Although their success is not yet dazzling (possibly because they are limited to the most severe cases of treatment-resistant depression), the results are very promising.[107, 108]

---

[104] M.L. Phillips, W.C. Drevets, S.L. Rauch, R. Lane, *Neurobiology of Emotion Perception II: Implications for Major Psychiatric Disorders*, "Biological Psychiatry" 2003, vol. 54, pp. 515–528.

[105] D. Ongür, A.T Ferry, J.L. Price, *Architectonic Subdivision of the Human Orbital and Medial Prefrontal Cortex*, "The Journal of Comparative Neurology" 2003, vol. 460, pp. 425–449.

[106] W.C. Drevets, K. Gadde, K.R.R. Krishnan, *Neuroimaging Studies of Depression*, [in:] *The Neurobiological Foundation of Mental Illness*, eds. D.S. Charney, E.J. Nestler, B.S. Bunney, 2nd edition, Oxford University Press, New York 2004.

[107] W.K. Goodman, R.L. Alterman, *Deep Brain Stimulation for Intractable Psychiatric Disorders*, "Annual Review of Medicine" 2012, vol. 63, pp. 511–524.

[108] P. Mohr, M. Rodriguez, A. Slavíčková, J. Hanka, *The Application of Vagus Nerve Stimulation and Deep Brain Stimulation in Depression*, "Neuropsychobiology" 2011, vol. 64, pp. 170–181.

Dominika Dudek, Rafał Jaeschke, Marcin Siwek
Jagiellonian University

The Moral Brain
Is There a Link between Neurobiology and Morality?[1]

> This is beyond our understanding, for, while we are all made of the same all-too-human substance, the same flesh, bones, blood, skin and laughter, tears and sweat, some of us become cowards and others heroes, some are aggressive and others passive.
>
> *José Saramago*

Introduction

The ongoing, stubborn attempts to close the gap between 'the universe of psyche' and 'the realm of matter' were among the hallmarks of the 20th century science. Paradoxically enough, some of the early calls for the urgent need to develop the empirical basis for psychology were formulated by Sigmund Freud and his fellow psychoanalysts – the Popperian heralds of the unfalsifiable, thus: the unscientific branch of knowledge.[2] The history of one of Freud's early works, *Project for a Scientific Psychology* (1895), is a recording of a failed effort to establish an evidence-driven neurobiological basis for modern psychology. Prior to the decision to abandon work on the manuscript,

---

[1] This paper was written within the research grant "The Limits of Scientific Explanation" sponsored by the John Templeton Foundation.

[2] Cf. K.R. Popper, *Droga do wiedzy. Domysły i refutacje*, translated by S. Amsterdamski, Wydawnictwo Naukowe PWN, Warszawa 1999, pp. 62–65.

Freud formulated some astonishingly shrewd hypotheses. In his opinion, what we experience as the conscious or unconscious processing of information is reflected in the neuronal architecture of the brain and the nervous system in general. As the neural connections are plastic in nature, they can be rearranged by numerous environmental factors – including the 'speaking therapy' that, in turn, might have the potential to change the character of psychological experiences.[3]

Freud's huge frustration over the outcomes of his work (notably, the work that had led to some conclusions which were very close to the contemporary understanding of the neurobiological basis of psychotherapy), resulting in his willingness to destroy the manuscript,[4] sounds like a confirmation of the cliché saying that nobody is a prophet in their own (scientific) home. The seed, however, had been planted, and the path to the new paradigm was opened.

While discussing the clinical aspects of the 'mind-and-matter relationships', it is worth mentioning the opinion expressed by Gin S. Malhi and Jim Lagopoulos, saying that the history of the 20th century's psychiatry had also been marked by attempts to define itself by 'linking the mind to the brain and developing a medical model for the study of psychological disorders'.[5] The climax of this trend came in 1998, when the Nobel Prize winner Eric Kandel put forward the following five-point 'intellectual manifesto', conceptualizing a common framework for psychiatry and the neuroscience:

> **Principle 1.** All mental processes, even the most complex psychological processes, derive from operations of the brain. The central tenet of this view is that what we commonly call mind is a range of func-

[3] See S. Freud, *Project for a Scientific Psychology* (1895), The Standard Edition of the Complete Psychological Works of Sigmund Freud, vol. 1, Hogarth Press, London 1953, pp. 283–397.

[4] See D. Centonze, A. Siracusano, P. Calabresi, G. Bernardi, *The Project for the Scientific Psychology (1895): A Freudian Anticipation of LTP-memory Connection Theory*, "Brain Reserch Reviews" 2004, vol. 46, pp. 310–314.

[5] See G.S. Malhi, J. Lagopoulos, *Making Sense of Neuroimaging in Psychiatry*, "Acta Psychiatrica Scandinavica" 2008, vol. 117, pp. 100–117.

tions carried out by the brain. The actions of the brain underlie not only relatively simple motor behaviors, such as walking and eating, but all of the complex cognitive actions, conscious and unconscious, that we associate with specifically human behavior, such as thinking, speaking, and creating works of literature, music, and art. As a corollary, behavioral disorders that characterize psychiatric illness are disturbances of brain function, even in those cases where the causes of the disturbances are clearly environmental in origin.

Principle 2. Genes and their protein products are important determinants of the pattern of interconnections between neurons in the brain and the details of their functioning. Genes, and specifically combinations of genes, therefore exert a significant control over behavior. As a corollary, one component contributing to the development of major mental illnesses is genetic.

Principle 3. Altered genes do not, by themselves, explain all of the variance of a given major mental illness. Social or developmental factors also contribute very importantly. Just as combinations of genes contribute to behavior, including social behavior, so can behavior and social factors exert actions on the brain by feeding back upon it to modify the expression of genes and thus the function of nerve cells. Learning, including learning that results in dysfunctional behavior, produces alterations in gene expression. Thus all of "nurture" is ultimately expressed as "nature".

Principle 4. Alterations in gene expression induced by learning give rise to changes in patterns of neuronal connections. These changes not only contribute to the biological basis of individuality but presumably are responsible for initiating and maintaining abnormalities of behavior that are induced by social contingencies.

Principle 5. Insofar as psychotherapy or counseling is effective and produces long-term changes in behavior, it presumably does so through learning, by producing changes in gene expression that alter the strength of synaptic connections and structural changes that alter the anatomical pattern of interconnections between nerve cells of the brain. As the resolution of brain imaging increases, it should eventually permit quantitative evaluation of the outcome of psychotherapy.[6]

Having translated Francis Crick's literary buzzword of 'the astonishing hypothesis'[7] – linking each and every mental phenomenon to neural activity – into a precisely defined scientific paradigm, Kandel provided the contemporary neuroscientists with useful armamentarium, making it possible (in theory, at least) to unweave the complexity of human behaviour and emotions. Alongside the development of modern diagnostic tools, an increasing number of issues strongly related to the humanities (including the question of morality) have been increasingly prone to empirical testing.

1. The Good, the Evil, and the Brain. Where is the Starting Point?

Possibly one of the most serious obstacles in discussing the neural background of morality is bound to the problem of choosing basic presumptions. As the beginnings of the history of Western ethics have usually been attributed to Socrates,[8] and the critical reflection over defining and justifying norms of dealing with the self and the others[9] has been present in the European culture ever after, it would be easy for a 'naturalist' to fall into the trap of a reductionistic bias. Facing such an impressive heritage, any 'reasonably simple' (i.e. the one narrow enough to escape Occam's razor) description of morality and related terms might be judged as dubious. Nevertheless, an attempt to

---

[6] E.R. Kandel, *The New Intellectual Framework for Psychiatry*, "American Journal of Psychiatry" 1998, vol. 155, pp. 457–469.

[7] See F. Crick, *The Astonishing Hypothesis. The Scientific Search for the Soul*, Touchstone, New York 1995.

[8] See R. Parry, *Ancient Ethical Theory*; [in:] *Stanford Encyclopedia of Philosophy*, ed. E. Zalta, < http://plato.stanford.edu/archives/fall2009/entries/ethics-ancient/>.

[9] See *Etyka* [in:] *Słownik filozofii*, ed. J. Hartman, Krakowskie Wydawnictwo Naukowe, Kraków 2009, p. 54.

delineate the boundaries of the key notions must be made. In other words: the question of the working definitions of good and evil should be addressed in order to define the possible starting point of the discussion. Afterwards, we will make an attempt to answer the question of whether there are any bridges between the 'conscious' understanding of ethical rules and the way our brains process moral dilemmas.

Morality is usually defined as a set of rules, ideas, values, judgments, opinions, and convictions existing in a given society, that regulate relationships between individuals and society as a whole, as well as between subjects themselves. Of note, the phenomena of real-life morality may differ substantially from the norms derived from theoretical (i.e. ethical) considerations. Consequently, the moral valence of a deed will be heavily influenced by cultural or situational factors.[10, 11] Interestingly enough, it would be an exaggeration to say that the distinction between good and evil has always been at the hub of the cultural processes over the history of humankind. According to Zygmunt Bauman,[12] the concept of morality, being separated from the entirety of thoughts, feelings and activities, is an idea very much rooted in modernity. In the pre-modern era, categories related to human proceedings such as 'usefulness', 'truth', 'beauty', or 'decency', had been largely irrelevant. As 'traditional' lifestyle (typical of societies of high stratification and stability provided by a dominant religious culture and roughly constant hierarchy of power relations[13]) has rarely been a subject of reflection for people immersed into that kind of reality, everything what had been happening seemed to be *of the same gravity and suitable for measure with the same scale.* The construct of 'free will' had been confined to the freedom of choosing

---

[10] See *Moralność*, [in:] *Słownik...*, *op. cit.*, pp. 142–143.

[11] Cf. H. Muszyński, *Rozwój moralny*, Wydawnictwa Szkolne i Pedagogiczne, Warszawa 1983.

[12] See Z. Bauman, *Etyka ponowoczesna*, translated by J. Bauman and J. Tokarska-Bakir, Wydawnictwo Naukowe PWN, Warszawa 1996, p. 9.

[13] See S. Seidman, *Zygmunt Bauman's Sociology of Postmodernity*, [in:] S. Seidman, *Contested Knowledge. Social Theory Today*, Blackwell Publishing, Maiden – Oxford – Carlton 2004, p. 193.

the evil over the good, i.e.: the (dubious) freedom of breaking the long-established habits of behaviour. Ultimately, 'choosing the path of the good' had not been a matter of free choice, but the opposite: it had unchangeably been the fruit of clinging to tradition.[14] The highly tempting question is, whether the neural correlates of moral decision--making coming down to following stiffly defined set of rules differ significantly from the neurobiological basis of solving moral challenges in the realm of the postmodern multiplicity of choices. Yet, the answer probably lies beyond the scope of contemporary neuroscience.

Going further: how could evil be defined? Philip Zimbardo has proposed the psychological concept of evil as 'intentionally behaving in ways that harm, abuse, demean, dehumanize, or destroy innocent others – or using one's authority and systemic power to encourage or permit others to do so on your behalf. In short, it is "knowing better but doing worse".[15] In view of such a broad spectrum of vicious activities, the narrowness of line between good and evil appears to be obvious, and on many occasions – referring to Hannah Arendt's famous diagnosis[16] – crossing it is a banal (not to be confused with unconscious) thing to do. "The exquisite nature of the human mind, which can create great works of art, science, and philosophy... [can also be] perverted to engage in acts of 'creative cruelty'".[17]

Following these introductory considerations, the vital question arises: where is the place for neural phenomena in our understanding of moral behaviour? Or (on more general level): can neuroscience provide us (in theory, at least) with some sort of ultimate explanation of morally valent activities?

The affirmative answer to the latter issue would probably be overoptimistic, as contemporary biomedical sciences and social psychol-

---

[14] *Ibid.*

[15] See P. Zimbardo, *The Lucifer Effect. Understanding How Good People Turn Evil*, Random House, New York 2007, p. 5.

[16] Cf. H. Arendt, *Eichmann in Jerusalem. A Report on The Banality of Evil*, The Viking Press, New York 1964.

[17] See P. Zimbardo, *The Lucifer..., op. cit.*, p. 9.

ogy tend to analyse the roots of good/evil dichotomy on the basis of different, complementary yet incompatible, paradigms (sounding like the distant echo of Charles P. Snow's idea of naturalists and humanists as members of 'the two cultures', who can never meet together on the intellectual grounds).[18] The first one – the dispositional model – operates on internal attributions, i.e. it is based on the assumption that the sources of the subject's vicious or heroic deeds are hidden within. In other words: individual responsibility derives from individual features (e.g. genetic predispositions, personality traits, or the subject's ability to steer the 'free will'). On the contrary, the situational model (preferred by social psychologists) refrains from internal attributions. This theoretical approach is concentrated on the network of circumstances that might have lead to the given outcome.[19] While the dispositional model is the predominant paradigm of contemporary neurobiology, psychiatry (as well as medicine in general), and related disciplines, it should be noted that pathogenesis of various mental disorders is explained by the template of the stress-vulnerability model.[20] The empirical usefulness of such a holistic approach seems to suggest that the unification of both models is not impossible.

Before we go into the detailed presentation of up-to-date evidence on the 'neurobiology of morality', an important remark on the two separate kinds of moral thinking should be emphasized. As moral decision-making cannot be reduced to following rational rules of conduct, it would be reasonable to assume that moral behaviour is also a function of emotions. In accordance with this presumption, moral violations can be divided into the 'personal' (involving the 'ME HURT YOU' pattern[21]; i.e. those kinds of abuse that are: "1) likely to

---

[18] Cf. C.P. Snow, *The Two Cultures and the Scientific Revolution*, Cambridge University Press, Cambridge – New York 1959.

[19] See P. Zimbardo, *The Lucifer…*, *op. cit.*, pp. 7–9.

[20] See e.g. R. van Winkel, N.C. Stefanis, J. Myin-Germeys, *Psychosocial stress and psychosis. A Review of the Neurobiological Mechanisms and the Evidence for Gene--stress Interaction*, "Schizophrenia Bulletin" 2008, vol. 34, pp. 1095–1105.

[21] See J. Greene, J. Haidt, *How (and Where) Does Moral Judgment Work?*, "Trends in Cognitive Science" 2002, vol. 6 (12), pp. 517–523.

cause serious bodily harm, 2) [done] to a particular person, 3) in such a way that the harm does not result from the deflection of an existing threat onto a different party"[22]) and 'impersonal' ones (i.e. all those types of violations that do not meet the criteria above).

Perhaps the most famous illustration of this distinction is the ethical thought experiment known as the trolley problem (originally formulated by Philippa Foot[23]): imagine yourself as the driver of a runaway vehicle that is about to run over and kill five people. However, there is a glimpse of 'hope': as you can hit a switch that will divert the trolley onto a different track where it will kill only one person instead of five. The impersonal moral problem would be: is pushing a button a right thing to do? Judith J. Thompson modified this scenario, turning an impersonal dilemma into a personal one:

> As before, a trolley is hurtling down a track towards five people. You are on a bridge under which it will pass, and you can stop it by dropping a heavy weight in front of it. As it happens, there is a very fat man next to you – your only way to stop the trolley is to push him over the bridge and onto the track, killing him to save five.[24]

Most research has shown that the majority of people tend to choose the 'lesser evil' (i.e. killing one person instead of five) in the first case, and 'letting the events go' in the second one (implying survivor of an obese individual, and the death of five anonymous subjects). According to Joshua Greene and Jonathan Haidt,[25] the possible explanation of such a distribution of answers (that are both very hard to justify, at least on a rational basis) can be enucleated from the fact that the 'fat man dilemma' includes the ME – HURT – YOU triad, while the im-

---

[22] See J.D. Greene, R.B. Sommerville, L.E. Nystrom, J.M. Darley, J.D. Cohen, *An fMRI Investigation of Emotional Engagement in Moral Judgment*, "Science" 2001, vol. 293, pp. 2105–2108.

[23] See P. Foot, *The Problem of Abortion and the Doctrine of the Double Effect in Virtues and Vices*, Basil Blackwell, Oxford 1978.

[24] See J.J. Thompson, *The Trolley Problem*, "Yale Law Journal" 1985, vol. 94, p. 1402.

[25] See J. Greene, J. Haidt, *How…*, *op. cit.*

personal problem of a runaway trolley lacks associations of this kind. In the Thompson's case harm that could be possibly done to an individual is so straightforward, that it fulfils the definition of 'HURT' put forward by Greene's research team (*the kinds of harm that a chimp can understand [e.g. assault, as opposed to, say, tax evasion]*).[26] Apparently the 'YOU' condition is also met, as an obese individual was clearly identified as a person – as a representation of Levinasian 'face of the Other'.[27] The same can be said about the 'ME' element, as a killer of an obese man would be the 'author' of the deed from start to finish, rather than a 'blind sword' operated by an external force. If one or more of the conditions are not met, then the moral dilemma would probably be impersonal, making 'radical solutions' much more likely. This finding sheds some interesting light on the issue of the apparent lack of remorse exhibited by air force drone operators[28] or – *toutes proportions gardées* – the clerks of Adolf Eichmann's Reich Security Main Office sub-department for Jewish affairs.[29]

From the neuroscientific point of view, the most important difference between personal and impersonal moral decisions lies in the fact that in the processing of the latter ones the rational assessment of available decisions' outcomes is predominant. On the contrary, these are emotions that seem to play a crucial role in the processing of personal dilemmas. Thus, personal or impersonal moral decisions seem to be mediated by different neural circuits. Having said this, we can finally progress to the discussion on the empirical data linking moral phenomena to various patterns of brain activity.

---

26 See J.D. Greene, R.B. Sommerville, L.E. Nystrom, J.M. Darley, J.D. Cohen, *An fMRI…*, *op. cit.*

27 See S. Žižek, *W obronie przegranych spraw*, translated by J. Kutyła, Wydawnictwo Krytyki Politycznej, Warszawa 2008, p. 298.

28 See P. Troop, *Killing With Drones, Proportionality, and Trolley Problems*, 28.10.2011, Practical Ethics Blog – University of Oxford. Ethics in News: http://blog.practicalethics.ox.ac.uk/2011/10/killing-with-drones-proportionality-and-trolley-problems/.

29 Cf. H. Arendt, *Eichmann…*, *op. cit.*

2. 'The Moral Deformity': The Neuropathology of Vice

It is no surprise that some early ideas linking distortions in moral judgment to brain abnormalities came as a result of the observations of individuals who had suffered brain damage.

As one man's poison is another man's meat, the milestone in research on the functional roles played by different brain regions were the consequences of the accident suffered by the 19th century American railroad foreman, named Phineas Gage.[30] Gage fell victim to the incident that took place in Cavendish, Vermont, on September 1848, when the premature blast of explosives used on a railway construction site left him with an iron rod (1.1 m long, 6 mm in diameter, and 6 kg in weight[31]) stuck in his head, running from the left cheek up to the vault of the scalp. Despite such a dramatic injury, he remained conscious and in relatively good condition. What made this case a historic one, however, happened a few weeks later: namely, a staggering change in Phineas' personality and intellectual capacity was observed. Dr John Harlow (the physician who was taking care of Gage) described him as being 'intellectually feeble', 'exceedingly capricious', 'childish', coarse, very impulsive, and indifferent towards others. "A child in his intellectual capacity and manifestations, he has the animal passions of a strong man"[32] – Harlow noted. Such a transformation of the man, once admired for his professional and social virtues, finally led to him being fired from the job, with subsequent social deterioration.

Three decades following Gage's accident, David Ferrier presented the earliest experimental evidence (derived from studies performed on monkeys), suggesting that consecutive areas of the cerebral cortex differ between each other in terms of function, and the destruction of

---

[30] See K. O'Driscoll, J.P. Leach, *'No longer Gage': An Iron Bar Through the Head. Early Observation of Personality Change After Injury to the Prefrontal Cortex*, "British Medical Journal" 1998, vol. 317, pp. 1673–1674.

[31] Cf. J.M. Harlow, *Passage of an Iron Rod Through the Head*, "Boston Medical and Surgical Journal" 1848, vol. 39, pp. 389–393.

[32] Cf. J.M. Harlow, *Recovery From the Passage of an Iron Bar Through the Head*, "Publications of the Massachusetts Medical Society" 1868, vol. 2, pp. 327–347.

prefrontal lobes – while exerting virtually no impact on neurological performance – causes "very decided alterations in the animal's character and behaviour (...) while not actually deprived of intelligence, they had lost, to all appearance, the faculty of the attentive and intelligence observation".[33] In a sense, Phineas Gage's fate was an unwanted and, by all means, tragic outpost in a brand new field of science.

Contemporarily, much of researchers' attention is being paid to the issue of personality changes (including those relevant for moral decision-making) taking place in subjects with certain neuropsychiatric diseases. Possibly the most convincing data on this problem derive from the studies regarding subjects with frontotemporal dementia (FTD), and its various subtypes (including Pick's disease[34]).

FTD belongs to the group of neurodegenerative disorders, characterized by a chronic clinical course, with progressive deterioration of cognitive functioning, altered behaviour, and personality changes, leading to severe disability, and – ultimately – death. In subjects with FTD neuropathological lesions are typically located in frontal lobes and anterior temporal regions, especially the ventromedial prefrontal cortex (vmPFC) and the orbitofrontal cortex (OFC).[35]

Compared to patients with Alzheimer's disease, subjects suffering from FTD tend to present with more prominent social and personality distortions, while cognitive and neuropsychological deficits are usually milder (on early stages of the illness, at least). Probably the most typical behavioural motifs observed in the course of FTD are related to difficulties in modulating social conduct. Notably, the following symptomatic triad: transgression of social norms (reflecting an impairment of the ability to modulate social behaviour), 'emotional blunting' (com-

---

[33] Cf. D. Ferrier, *The Goulstonian Lecutres on the Localisation of Cerebral Diseases*, "British Medical Journal" 1878, vol. 1, pp. 443–447.

[34] Cf. J.R. Hodges, R. Rhys-Davies, J.H. Xuereb, B. Casey, M. Broe, T.H. Bak, J.J. Krill, G.M. Halliday, *Clinicopathological Correlates in Frontotemporal Dementia*, "Annals of Neurology 2004", vol. 56, pp. 399–406.

[35] See H.J. Rosen, M.L. Gorno-Tempini, W.P. Goldman, R.J. Perry, N. Schnuff, M. Weiner, R. Feiwell, J.H. Kramer, B.L. Miller, *Patterns of Brain Atrophy in Frontotemporal Dementia and Semantic Dementia*, "Neurology" 2002, vol. 58, pp. 198–208.

ing down to a loss of empathy), and loss of insight for the subject's behaviours and their consequences constitute the core clinical features of the illness. Consequently, as the intensity of basic drives remain intact, a lack of empathy often leads to overtly sociopathic acts, such as sexual indiscretions, physical aggression, stealing, breaking into other people's properties, and paraphilias. The severity of those symptoms is even greater in subjects whose OFC is affected, as the presence of orbitofrontal lesions leads to the loss of impulse control (e.g. see the case report of a patient with a brain tumour who developed paedophilia[36]). Mario F. Mendez and colleagues[37] termed the essence of FTD a disturbance in the neurobiological substrate of morality. A subject suffering from the illness exhibit with "diminished emotional identification with others and [solving] moral dilemmas in an impersonal fashion, while having retained knowledge of moral rules and norms and [the ability of reasoning] about the right and wrong of a situation".[38]

Those changes, observed early in course of FTD, seem to be specifically related to the failure of vmPFC. This part of prefrontal cortex has multiple interconnections with the limbic system[39] (the cerebral region that is responsible for emotional processing[40]), therefore it "guides social behaviour by re-experiencing previously learned emotional responses and computing shared representations"[41] (exemplified by the situation when the emotional state of another person stimulates – possibly via the mirror neurons – the analogous emotional state in

[36] See J.M. Burnes, R.H. Swerdlow, *Right Orbitofrontal Tumor with Pedophilia Symptom and Constructional Apraxia Sign*, "Archives of Neurology" 2003, vol. 60, pp. 437–440.

[37] See M.F. Mendez, E. Anderson, J.S. Shapira, *An Investigation of Moral Judgement in Frontotemporal Dementia*, "Cognitive and Behavioral Neurology" 2005, vol. 18, pp. 193–197.

[38] *Ibid.*

[39] See D. Tranel, *'Acquires sociopathy': The Development of Sociopathic Behawior Following Focal Brain Damage*, "Progress in Experimental Personality & Psychopathology Research" 1994, vol. 17, pp. 285–311.

[40] Cf. Sadock B.J., Sadock V.A. (eds.), *Kaplan & Sadock's Synopsis of Psychiatry. Behavioral Sciences/Clinical Psychiatry. Tenth Edition*, Wolters Kluwer, Lippincott Williams & Wilkins 2007, pp. 89–90.

[41] See M.F. Mendez, E. Anderson, J.S. Shapira, *An Investigation..., op. cit.*

the given subject). Accordingly, patients with impaired vmPFC function will probably be less empathetic and concerned about others. The research on subjects with FTD in whom other brain regions were involved, suggest that there is some level of differentiation within sociopathic traits depending on the location of the challenged cerebral structure. M.F. Mendez noted that FTD patients who had their right anterior temporal lobes affected were characterized by interpersonal coldness and a lack of responsiveness to other's distress, and damage to the right ventromedial-anterior temporal network led to decreased emotional empathy and reduced responsiveness to victims.[42] The 'region-specificity' of such a pattern of disturbed moral emotions and behaviours has been confirmed in studies of non-FTD patients with lesions (e.g. due to a tumour, injury, or anterior cerebral artery aneurysm) located in the vmPFC. Thus, having compared 7 subjects with vmPFC damage and 12 healthy controls, Elisa Ciaramelli et al. noted that the patients had been more likely to accept personal moral violations that maximize beneficial consequences (exemplifying utilitarian moral judgments), while no significant differences were observed in terms of solving impersonal moral dilemmas or non-moral issues.[43] The members of Michael Koenigs' research team have also confirmed this result.[44] Furthermore, Giovanna Moretto et al. have presented evidence suggesting that the core function of vmPFC with regard to moral behaviour is mediating the anticipation of emotional consequences of personal moral abuses.[45] This finding puts an important question on the table: is it merely the potential to experience remorse which makes our brains 'moral' ones?

---

[42] See M.F. Mendez, *The Unique Predisposition to Criminal Violations in Frontotemporal Dementia*, "Journal of the American Academy of Psychiatry and the Law" 2010, vol. 38, pp. 318–323.

[43] See E. Ciaramelli, M. Muccioli, E. Làdavas, G. Di Pellegrino, *Selective Deficit in Personal Moral Judgment Following Damage to Ventromedial Prefrontal Cortex*, "SCAN" 2007, vol. 2, pp. 84–92.

[44] See M. Koenigs, L. Young, R. Adolphs, D. Tranel, F. Cushman, M. Hauser, A. Damasio, *Damage to the Prefrontal Cortex Increases Utilitarian Moral Judgments*, "Nature" 2007, vol. 446, pp. 908–911.

[45] See G. Moretto, E. Làdavas, F. Mattioli, G. Di Pellegrino, *A Psychophysiological Investigation of Moral Judgment after Ventromedial Prefrontal Damage*, "Journal of Cognitive Neuroscience" 2010, vol. 22, pp. 1888–1899.

3. Beyond Neuropathology: Moral Lessons from Psychopathy

Historically, the term of psychopathy was used to name all possible kinds of personality disorders,[46] making the notion highly unclear. Ultimately, its meaning was constrained to such a kind of personality disturbance that involves the triad of characteristic dimensions: "an arrogant and deceitful style of interpersonal functioning, a deficiency in experiencing, understanding and expressing emotions, and highly impulsive behaviour".[47]

Contemporarily, the negatively associated term of psychopathy is not used in the literature anymore; instead the name of antisocial or dissocial personality disorders is employed. While the 10th Edition of the World Health Organization's International Classification of Diseases[48] contains an in-depth definition of dissocial personality disorder (see Box 1), the historical notion has not been fully abandoned, as some authors strongly advocate the appropriateness of the word 'psychopathy'. Accordingly, Łukasz Barwiński and Józef K. Gierowski[49] point out the fact that psychopathy is more about the fundamental distortion of the personality structure (as expressed by disorders challenging the emotional and interpersonal sphere of human functioning) than antisocial behaviours by themselves. Therefore, the term of 'dissocial personality disorder' may be both misleading and too narrow for the problem.

---

[46] Cf. A. Kępiński, *Psychopatie*, Wydawnictwo Literackie, Kraków 2002.

[47] See L. Johnstone, D.J. Cooke, *Psychopathic-like Traits in Childhood: Conceptual and Measurement Concerns*, "Behavioral Science & the Law" 2004, vol. 22, pp. 103–125.

[48] See World Health Organization, *The ICD-10 Classification of Mental and Behavioural Disorders. Diagnostic Criteria for Research*, <http://www.who.int/classifications/icd/en/GRNBOOK.pdf>.

[49] See Ł. Barwiński, J.K. Gierowski, *Controversies Around the Concept of Psychopathy and Its Usefulness in Forensic Psychological Reporting Concerning Adolescents*, "Problems of Forensic Sciences" 2008, vol. 78, pp. 382–402.

Box 1. The general definition of personality disorders and the detailed characteristics of dissocial personality disorder (according to International Classification of Diseases, 10th Edition)

Specific personality disorders

G1. Evidence that the individual's characteristic and enduring patterns of inner experience and behaviour deviate markedly as a whole from the culturally expected and accepted range (or 'norm'). Such deviation must be manifest in more than one of the following areas:

(1) cognition (i.e. ways of perceiving and interpreting things, people and events; forming attitudes and images of self and others);

(2) affectivity (range, intensity and appropriateness of emotional arousal and response);

(3) control over impulses and need gratification;

(4) relating to others and manner of handling interpersonal situations.

G2. The deviation must manifest itself pervasively as behaviour that is inflexible, maladaptive, or otherwise dysfunctional across a broad range of personal and social situations (i.e. not being limited to one specific 'triggering' stimulus or situation).

G3. There is personal distress, or adverse impact on the social environment, or both, clearly attributable to the behaviour referred to under G2.

G4. There must be evidence that the deviation is stable and of long duration, having its onset in late childhood or adolescence.

G5. The deviation cannot be explained as a manifestation or consequence of other adult mental disorders, although episodic or chronic conditions from sections F0 to F7 of this classification may co-exist, or be superimposed on it.

G6. Organic brain disease, injury, or dysfunction must be excluded as possible cause of the deviation (if such organic causation is demonstrable, use category F07).

Dissocial personality disorder
A. The general criteria of personality disorder (F60) must be met.

B. At least three of the following must be present:

(1) Callous unconcern for the feelings of others.

(2) Gross and persistent attitude of irresponsibility and disregard for social norms, rules, and obligations.

(3) Incapacity to maintain enduring relationships, though having no difficulty to establish them.

(4) Very low tolerance to frustration and a low threshold for discharge of aggression, including violence.

(5) Incapacity to experience guilt, or to profit from adverse experience, particularly punishment.

(6) Marked proneness to blame others, or to offer plausible rationalizations for the behaviour bringing the subject into conflict with society.

Comments: Persistent irritability and the presence of conduct disorder during childhood and adolescence, complete the clinical picture but are not required for the diagnosis.

Notably, the emotional and behavioural features of psychopathy are similar to those observed in subjects with prefrontal cortex damage. Thus, the question arises as to whether psychopathic traits can be linked to the dysfunction of analogous cerebral structures to the ones described in the previous chapter, or is it just a coincidental resemblance?

Having reviewed current neuroimaging data, Yu Gao and co-authors came to the conclusion that – in comparison with the general population – psychopaths tend to exhibit prefrontal cortex alterations. Some research using voxel-based morphometry has revealed that the population discussed is characterized by significant grey matter volume reduction in the prefrontal cortex (as reflected by the reduced frontotemporal perfusion, noted in studies utilizing single-photon emission computed tomography). Interestingly enough, the prefrontal cortex does not seem to be the only affected part of the brain. Accordingly, the size of the posterior hippocampus has been found to be reduced in antisocial alcohol-dependent subjects with high psychopathic scores. Volume reductions in the bilateral amygdala (particularly the basolateral and superficial nuclei groups), and increased callosal white matter volume have also been noted. From the functional point of view, psychopathy seems to be related to an inappropriate pattern of activation in the frontal-temporal circuit. While performing semantic tasks, psychopaths do not exhibit correct neural differentiation between abstract and concrete stimuli in the right superior temporal gyrus, left ventrolateral prefrontal cortex, middle temporal cortex, and anterior cingulate cortex. In subjects with psychopathic personality traits, the dorsolateral and ventrolateral prefrontal cortex, as well as the anterior and posterior cingulate cortex, amygdala, hippocampus, and parahippocampus gyrus fail to respond adequately to affect-related activation.[50]

What can be made of these data? Gao et al. have drawn the conclusion that while there is no single 'psychopathy centre' in the brain, various cerebral regions are responsible for the discrete features of the disorder. Accordingly, impulsivity and impaired moral judgment are probably related to functional and/or anatomical deficits of structures responsible for decision-making, behavioural steering and emotional control (i.e. the orbitofrontal, ventromedial prefrontal, and the cingulate cortex). On the other hand, shallow affect and lack of empathy both seem to be the outcomes of dysfunctions of medial temporal regions (particularly the hippocampus and amygdala), as they are the key elements of the 'emotional processing system'. Addressing the issue of the pathogenesis of psychopathy, the cited authors have put forward a hypothesis linking psychopathic personality traits to some basic

[50] See Y. Gao, A.L. Glenn, R.A. Schug, Y. Yang, A. Raine, *The Neurobiology of Psychopathy: A Neurodevelopmental Perspective*, "Canadian Journal of Psychiatry" 2009, vol. 54, pp. 813–823.

neurodevelopmental distortions of the cerebral structure and functioning.[51] Andrea L. Glenn and colleagues' opinion is also worth mentioning, as they consider the amygdala to be at the hub of the pathology leading to psychopathy. The amygdala is the crucial structure in determining responses to signals indicating distress in others; therefore its dysfunction leads to antisocial behaviours, as well as it seems to determine the susceptibility towards solving personal moral dilemmas in a utilitarian manner.[52] Consequently, a 'psychopath' can be defined as somebody who knows right from wrong but who does not care.[53]

4. Returning to the 'Moral Majority': What Can Be Said about the Neurobiological Substrates of Morality in Healthy Subjects?

From the methodological standpoint, research on the neural correlates of moral behaviour in healthy individuals do not differ substantially from the trials engaging subjects with neuropsychiatric illnesses or disorders. Accordingly, in most of the studies participants have been encouraged to perform some tasks regarding dilemmas of moral judgment or reasoning, and, while involved in such processings, their neural activities have been analyzed with functional magnetic resonance imaging (fMRI).[54]

In 2002 Jorge Moll and colleagues published the far-reaching results of the study,[55] aiming to differentiate between specific cerebral

[51] *Ibid.*

[52] See A.L. Glenn, A. Raine, R.A. Schug, *The Neural Correlates of Moral Decision--making in Psychopathy*, "Molecular Psychiatry" 2009, vol. 14, pp. 5–6.

[53] See M. Cima, F. Tonnaer, M.D. Hauser, *Psychopaths Know Right From Wrong But Don't Care*, "SCAN" 2010, vol. 5, pp. 59–67.

[54] See M.F. Mendez, *The Neurobiology of Moral Behavior: Review and Neuropsychiatric Implications*, "CNS Spectrums" 2009, vol. 14, pp. 608–620.

[55] See J. Moll, R. de Oliveira-Souza, P.J. Eslinger, I.E. Bramati, J. Mourão-Miranda, P.A. Andreiuolo, L. Pessoa, *The Neural Correlates of Moral Sensitivity: A Functional Magnetic Resonance Imaging Investigation of Basic and Moral Emotions*, "Journal of Neuroscience" 2002, vol. 22(7), pp. 2730–2736.

structures involved in the processing of morally valent or non-valent stimuli. Therefore, participants underwent fMRI diagnostics at the time of watching either emotionally charged scenes with or without moral content, or emotionally neutral pictures. The investigators found that the essential moral-processing structures were: right medial orbitofrontal cortex, medial frontal gyrus, and the cortex surrounding the right posterior superior temporal sulcus – i.e. the places in the brain that were activated while watching morally charged scenes. A year earlier, Joshua Greene et al.[56] presented evidence suggesting that personal or impersonal moral violations imply different patterns of neural activity. They found that subjects who committed personal abuses exhibit higher activity of the medial frontal gyrus, posterior cingulate gyrus, superior temporal sulcus, and inferior parietal lobule. On the contrary, impersonal moral transgressions implied higher activity of the vmPFC and parietal lobe.

In an attempt to summarize the state-of-the-art knowledge on the 'moral brain', Mario F. Mendez[57] pointed out the functions of the crucial elements of the system described. Accordingly, the vmPFC plays a pivotal role in the neuromoral network, as it becomes activated with tasks requiring explicit moral judgments, passive viewing of morally salient photos, and the elicitation of charity, fairness, guilt, and other moral emotions and sentiments. It also seems to be the driving force behind solving personal moral dilemmas, due to its profound involvement in anticipating somebody else's harm (therefore, promoting 'prosocial' reactions). Of note, in subjects who do not have sufficiently robust vmPFC activity, the dorsolateral prefrontal cortex gains a relative functional advantage, implying a tendency towards solving moral problems in a reasoned (impersonal) way. The region of the orbitofrontal and ventrolateral cortex (OFC/VL) seems to be responsible for the mediation of socially aversive responses, as well as

---

[56] See J.D. Greene, R.B. Sommerville, L.E. Nystrom, J.M. Darley, J.D. Cohen, *An fMRI…*, *op. cit.*

[57] See M.F. Mendez, *The Neurobiology…*, *op. cit.*

inhibition of impulsive/automatic responses (as opposed to the role of the amygdalae, mediating response to threat and aversive social and moral learning). OFC/VL and the nearby anterior insula and amygdala may also play an additional role to the function of the vmPFC, as they promote altruistic punishments (dependent on determining that others, particularly those with a negative reputation, are deliberately not playing by the rules) driven by sentiments related to social aversion or exclusion (e.g. anger, indignation, disgust or contempt).

Conclusion

In spite of the enormous progress observed in the field of neuroscience, it would be premature to say that Freud's postulate of developing 'scientific psychology' has finally been fulfilled. While there has been quite a lot of stimulus-based studies (concerning relatively simple phenomena), the research that associates patterns of neural activity with behaviour are still a relative novelty. When it comes to the issue of morality, the understanding of the neurobiological background of moral decision-making is still rather crude, as the relevant findings have been observed on the level of only gross cerebral structures.[58] Even though some of the crucial elements of the 'moral brain' have been identified, the question about the influence of the premorbid personality (i.e. whether somebody had been a 'good' or a 'bad' person) on the outcomes of the failure of neuromoral circuits is still to be answered.

Yet, regardless of all the reservations, we should strongly emphasize that the neural basis of moral behaviour is no longer a complete abstraction. The cornerstone of a new branch of science has definitely been founded, and there are lots of fascinating paths to follow.

---

[58] Cf. J.D. Greene, *The Cognitive Neuroscience of Moral Judgment*, online version, <http://www.wjh.harvard.edu/~jgreene/GreeneWJH/Greene-CogNeuroIV-09.pdf>.

Bram T. Heerebout, R. Hans Phaf
University of Amsterdam

Behavioural Benefits and Neural Costs in Simulated Evolution[1]

Introduction

"There is no such thing as a free lunch" is the title the economist Milton Friedman chose for his famous book in which he expresses the idea that one cannot get something for nothing.[2] Although often ignored, this is particularly true for the brain. The brain thirsts for energy; in humans consuming even 20 percent of the total metabolic cost of the resting body.[3] When explaining the prevalence of certain behaviours over others, it is mostly done in terms of greater benefits. The tremendous energy and other costs of the brain are often not considered. Nevertheless, in our evolutionary past, there must have been a strong selective pressure to constrain the brain's energy consumption. The neural costs of an adaptation are all too often ignored in evolutionary psychology. Computational methods of simulating evolution can restore the balance between behavioural benefits and neural costs in cognitive models.

---

[1] This paper was written within the research grant "The Limits of Scientific Explanation" sponsored by the John Templeton Foundation.

[2] M. Friedman, *There's No Such Thing as a Free Lunch: Essays on Public Policy*, Open Court Publishing Co., LaSalle 1975.

[3] J.W. Mink, R.J. Blumenschine, D.B. Adams, *Ratio of CNS to Body Metabolism in Vertebrates: a Demonstration of Its Constancy and Functional Basis*, "American Journal of Physiology" 1981, vol. 241, pp. 203–212.

Besides the enormous energy consumption, there are other natural laws that constrain the brain's development. Probably, the minimally required physical space that neurons and their interconnections occupy had a great impact and, additionally, the speed with which neurons can transfer signals and process information likely influenced the course of evolution. Particularly in urgent situations, time consumption determines the upper limit of the processing capacity of the brain. We will argue that the time constraints imposed on the brain play an essential part in the ultimate cause of behaviours, and can explain behaviour that intuitively seems maladaptive. Unexpected behaviour observed in feeding birds may serve as an illustrative example.[4] They spend an enormous amount of their energy hurrying to find food to bring back to the nest. It has been recorded that a parent with food in its beak was distracted along the way by a nest with unrelated young, signalling for food. The parent, apparently unable to resist the innate urge to feed, put the food in these mouths, even though it was not its own offspring. Typically, birds that feed their young are predisposed to react strongly to brightly coloured open beaks, but apparently do not process much other information concerning chicks. This can be explained by the constraining neural cost of distinguishing unrelated offspring from your own. It can be cheaper to occasionally make a mistake than to suffer the costs of more complicated neural tissue and never err.

The study of a behaviour's prior evolutionary development is problematic. The next section of this chapter describes how evolutionary psychologists study current behaviour to learn about its past. Their methods and conclusion are often controversial and the vast knowledge base from traditional psychology about the role of mental functions in individual and social behaviour seems readily ignored. Instead, a prevalent behaviour is assumed to have some adaptive value in an evolutionary environment, either now or in a past 'en-

---

[4] A.F. Skutch, *Helpers among Birds*, "The Condor" 1961, vol. 3, pp. 198–226.

vironment of evolutionary adaptation'.[5] Greiling and Buss, for instance, write about women's sexual strategies: "…it is unlikely that so pervasive a pattern would be pursued by women unless there were contexts in which it was advantageous and benefits sufficiently large to be reaped in those contexts".[6] We will argue that a computational method can provide insights that overcome the shortcomings associated with purely theoretical (i.e., "armchair") evolutionary reasoning, because it allows modellers to factor in the cost of an adaptation.

The shortcomings of evolutionary psychology may be addressed with computational modelling methods, which will be discussed in the subsequent section. The method of evolutionary computation extends traditional 'Genetic Algorithms', with agents in virtual environments controlled by artificial neural networks, and thus link behavioural to neural features.[7] The connections of the neural networks, encoded by genes, are subjected to a simulated 'natural selection' and genetic modification. The interactions between the genes and the neurons, the neurons and the resulting actions, and the action with the environment are all modelled as independent, local processes. This allows for the emergence of complex systems taking into account all sorts of constraining factors, such as the time and energy costs of neural processing. As the networks are derived only indirectly from modelled selection pressures, physical relations and constraints, the resulting global neural mechanisms can evolve in an analogous manner to mechanisms in biological brains. The data from this synthetic approach to evolutionary cognitive neuroscience can lead to novel insights into brain functions and mechanisms and contribute to a better understanding of mental processes.

---

[5] S.J. Gould, *Darwinian Fundamentalism, part 1*, "New York Review of Books" 1997, pp. 34–37.

[6] H. Greiling, D.M. Buss, *Women's Sexual Strategies: The Hidden Dimension of Extra-pair Mating*, "Personality & Individual Differences" 2000, vol. 28, p. 962.

[7] See R.D. Beer, *Intelligence as Adaptive Behavior: An Experiment in Computational Neuroethology*, Academic Press, St. Louis 1990; P. den Dulk, B.T. Heerebout, R.H. Phaf, *A Computational Study into the Evolution of Dual-route Dynamics for Affective Processing,* "Journal of Cognitive Neuroscience" 2003, vol. 15, pp. 194–208.

In the last section, we will illustrate this methodology with studies of the evolution of affective processing. When balancing behavioural benefits and neural costs in evolutionary simulations, processes that seem relevant to emotions and affect emerge naturally. Similarly to the bird feeding example, affective processes may at times be maladaptive, but on the whole endow the organism with large fitness-enhancing capabilities. Because these capabilities have been implemented by simple mechanisms with the lowest neural costs, they may overshoot in behaviour with no obvious advantage to the organism. We will even consider such an optimal balance of neural costs and behavioural benefits as an alternative explanation for altruistic behaviour.

1. Empirical Methods for the Study of Adaptive Behaviour

A full understanding of all current behaviour that is shared by all members of the human species requires knowledge about the evolution of the brain. Yet collecting empirical data for the study of the evolutionary history of mind and brain is extremely challenging. The mental processes of our ancestors, and the underlying neural circuitry, cannot be observed directly. Traditional approaches to study evolutionary development are problematic. The analysis of the fossil record is of no avail since brains do not fossilize. Over the past decade, the use of comparative methods for the investigation of the brain's history has certainly produced a vast body of knowledge.[8] It has revealed, for example, that the cortex is present in all mammalian taxa and that the earliest mammals already had a six-layered neo-cortex. Nevertheless, despite all recent advances, the comparative approach has left many questions unanswered. Because the function of specific neuronal structures may have shifted greatly over thousands or millions of years, constants in structure do not necessarily entail constants in function.

---

[8] Cf. R.G. Northcutt, J.H. Kaas, *The Emergence and Evolution of Mammalian Neocortex*, "Trends in Neurosciences" 1995, vol. 18, pp. 373–379.

Alternatively, evolutionary psychologists collect data on current behaviour in order to study its evolutionary origin. A core, but often tacit, assumption of their approach is that behaviour is ultimately a product of evolution. Central themes, studied in evolutionary psychology, are why we choose particular mates,[9] avoid incest,[10] and favour kin.[11]

Daly and Wilson's well-known work on the 'Cinderella effect' may serve as an example of such a study.[12] The main question was whether the favouring of kin also worked in the opposite direction by disfavouring stepchildren. They collected statistics on infanticide by stepparents and genetic parents, and indeed concluded that stepchildren are at a higher risk of maltreatment, particularly fatal beatings. They explained this by Hamilton's inclusive fitness theory, which argues that reproductive success is determined by one's own reproductive success combined with that of one's genetic relatives.[13] If an organism shares resources with genetic relatives, and especially with its offspring, then it also increases the odds of the reproduction of its own genes. For this reason, nurturing your offspring enhances your inclusive fitness while nurturing your stepchildren does not.

---

[9] E.g. D. Penn, W. Potts, *The Evolution of Mating Preferences and Major Histocompatibility Complex Genes*, "American Naturalist" 1999, vol. 153, pp. 145–164; H. Greiling, D.M. Buss, *op. cit.*

[10] E.g. A.P. Wolf, *Sexual Attraction and Childhood Association: A Chinese Brief for Edward Westermarck*, Stanford University Press, Stanford 1995; M.H. Stone, E. Steinmeyer, J. Dreher, M. Krischer, *Infanticide in Female Forensic Patients: The View From the Evolutionary Standpoint*, "Journal of Psychiatric Practice" 2005, vol. 11, pp. 35–45.

[11] E.g. H. Keller, *Human Parent-Child Relationships from an Evolutionary Perspective*, "American Behavioral Scientist" 2000, vol. 43, pp. 957–969; S. Platek, J.P. Keenan, F. Mohamed, *Sex Differences in the Neural Correlates of Child Facial Resemblance: An Event-related fmri Study*, "Neuroimage" 2005, vol. 25, pp. 1366–1344.

[12] M. Daly, M. Wilson, *Homicide*, Aldine de Gruyter, Hawthorne, NY 1988; M. Daly, M. Wilson, *An Assessment of Some Proposed Exceptions to the Phenomena of Nepotistic Discrimination Against Stepchildren*, "Annales Zoologici Fennici" 2001, vol. 38, pp. 287–296.

[13] W.D. Hamilton, *The Genetical Evolution of Social Behaviour I*, "Journal of Theoretical Biology" 1964, vol. 7, pp. 1–16; W.D. Hamilton, *The Genetical Evolution of Social Behaviour II*, "Journal of Theoretical Biology" 1964, vol. 7, pp. 17–52.

By integrating psychology within the evolutionary framework, an undesirable divide with the natural sciences may be eliminated. Miller and Todd wrote: "The physical and biological sciences take for granted a compatibility principle: chemistry must be consistent with physics, even if chemistry cannot (yet?) be derived from physics. Mutual compatibility between sciences leads to conceptual integration. Darwin's central importance in the history of science lies in his extension of the compatibility requirement to the biological realms: evolutionary theory provides the explanatory bridge between physical simplicity and functional organic and behavioral complexity".[14] In turn, evolutionary psychology aims at bridging the gap between biology and psychology.

1.2 Problems with Evolutionary Psychology

Studies in the field of evolutionary psychology have received much criticism, amongst others for its readiness to accept "just so" stories.[15] Critics claim a study will typically focus on one singled-out type of behaviour, assume it is an almost instantaneous adaptation and then investigate possible conditions that may have made the trait favourable. The fact that mechanisms may be far older than the human species is often ignored and reasoning is regularly too simplistic. The influence of complex dynamics between different behaviours is mostly neglected.

---

[14] G.F. Miller, P.M. Todd, *A Bottom-up Approach with a Clear View of the Top: How Human Evolutionary Psychology can Inform Adaptive Behavior Research*, [Review of the book *The Adapted Mind: Evolutionary Psychology and the Generation of Culture*], "Adaptive Behavior" 1994, vol. 3, p. 87.

[15] Cf. S.J. Gould, R.C. Lewontin, *The Spandrels of San Marcos and the Panglossian Program: A Critique of the Adaptationist Program*, "Proceedings of the Royal Society of London, Series B" 1979, vol. 205, pp. 581–598; E.A. Lloyd, M.W. Feldman, *Evolutionary Psychology: A View From Evolutionary Biology*, "Psychological Inquiry" 2002, vol. 2, pp. 150–156; J.J. Bolhuis, C.D.L. Wynne, *Can Evolution Explain How Minds Work?*, "Nature" 2009, vol. 458, pp. 832–833.

In general, genetic mutations studied independently of their genetic context can lead to faulty conclusions. Whether a particular mutation is favourable depends very much on the individual's other inheritable traits. For example, in ethology, the benefit of having the tendency for an extensive courting period is explained by the tendency to commit to long-term monogamous relationships. Probably, the impressive rituals some birds have to perform before entering the commitment raises the threshold for those individuals that have a preference for a short-term mating strategy. One trait, thus, can evolve to benefit another. Following this argument, a mutation that modifies the central nervous system, subsequently influencing a behavioural change, cannot be effectively interpreted independent of the individuals other genetic behavioural dispositions. Consequently, we err by singling out a behavioural tendency and simply reducing it to a universal fitness enhancing or diminishing trait. Its adaptive value is derived from the complex interactions and influences within a web of co-evolved behavioural tendencies.

The studies of Daly and Wilson[16] on infanticide and child abuse are often mentioned in reviews critical of evolutionary psychology.[17] Daly and Wilson only focus on an extremely limited aspect of the behaviours associated with the relation between child and caregiver. According to De Waal this is awkward because the vast majority of parents, regardless of whether they are stepparents or biological parents, lovingly care for their children.[18] Even if adopted children are more often the victim of abuse, by far, the majority of adopted children are well cared for. An evolutionary explanation for the behaviour of a minority of stepparents likely accounts for the majority as well. De Waal has called this 'the dilemma of the rarely exercised option';

---

[16] M. Daly, M. Wilson, *Homicide*, *op. cit.*; M. Daly, M. Wilson, *An Assessment...*, *op. cit.*

[17] Cf. D.J. Buller, *Evolutionary Psychology: The Emperor's New Paradigm*, "Trends in Cognitive Sciences" 2005, vol. 9, pp. 277–283; F.B.M. de Waal, *Evolutionary Psychology: The Wheat and the Chaff*, "Current Directions in Psychological Science" 2002, vol. 11, pp. 187–191.

[18] F.B.M. de Waal, *op. cit.*

a Darwinian explanation for an atypical behavioural choice is incomplete without an equally good justification for the typical choice.[19] It is even unclear whether infanticide by stepparents is best accounted for in evolutionary terms at all. The same reasons that may have influenced the replacement of parents by stepparents (e.g., marital conflict) may have increased the chances of violence towards the children. In addition, attachment relations will generally have had less time to develop between children and stepparents than between children and biological parents.

Most evolutionary psychologists do not investigate the link between behaviour and brain processes and do not specify the neural bases for their hypothesized innate behaviour. Caring behaviour, for instance, is most likely produced by the same neural structures in both biological and foster parents. Hence, evolutionary change will influence both type of parents. In this case, evolutionary accounts can thus not simply be constructed fully independently of each other. In our view, evolutionary-old behaviour results from evolutionary-prepared neural processes and structures. If it is unlikely that these exist, it is also unlikely that the behaviour has resulted from natural selection.

1.3 The Proximate and the Ultimate Levels of Analysis

Evolutionary Psychology seeks adaptive causes for specific behaviour. In traditional psychology only the immediate mechanisms that produce behaviour are studied. Typically, in the field of animal behaviour these two levels of analysis are explicitly distinguished: the 'proximate', which explains behaviour based on local trigger stimuli and underlying mechanisms, and the 'ultimate', which explains behaviour based on evolutionary adaptation from a global perspective. What is often missing from evolutionary psychologists reasoning is

[19] *Idem*, *Survival of the rapist*, [Review of the book *A Natural History of Rape: Biological Bases of Sexual Coercion*], "New York Times Book Review" 2000, pp. 24–25.

that these two levels can be further divided into what is now known as 'Tinbergen's four questions'. The proximate level encompasses both the 'mechanism' and the 'process of development' (i.e., questioning how an organism's structures work and how they develop during the individual's lifetime). Explanations at the ultimate level can be in terms of 'phylogeny' or in 'functional adaptation' (i.e., questioning the evolutionary history of a trait over many generations and why this trait would have enhanced reproduction or survival).

A simple example of explaining behaviour at the four levels may be why animals search for food. It is because their energy levels run low and this is somehow (in humans through a lowering of acidic value in the stomach) signalled to the brain (proximate – mechanistic). The type of food that is craved for can also be influenced by the age of the animal or its life phase (proximate – developmental). Of course, animals that have never felt the inkling that food was needed, had steeply decreased odds for reproduction and survival. Those animals that developed an attraction to food thus passed this trait to their offspring (ultimate – adaptation). The types of food that are naturally craved for may not necessarily always be very healthy for the animals. These foods may have become scarce or, conversely, have become overabundant. They were however, probably the types of food that increased fitness in the past (ultimate – phylogenetic). Either mechanism, taken apart from the whole of the environment and an organism's ecological interaction, fails to provide a systematic explanation at a global level.

Evolutionary psychology clearly addresses the ultimate level of explanation. It has been observed however that in doing so, the large body of knowledge from traditional psychology that addresses the proximate level, is too often neglected and the different levels are rarely linked together.[20] Insights into the proximate mechanisms of behaviour in current organisms will, of course, be crucial in understanding the value of the behaviour for the organisms' ancestors.

[20] F.B.M. de Waal, *Evolutionary Psychology...*, *op. cit.*

In essence, for evolutionary psychologists to understand the ultimate cause for a behaviour often boils down to deciding whether it has, or does not have, an adaptive value. This simple dichotomy ignores other possibilities, for example that a behaviour results from some malfunction. Or the reason may be a bit more complex. Consider the example mentioned earlier of the bird feeding unrelated offspring. It would seem a malfunction, but we argued that it may really constitute an adaptation which minimizes overall neural costs while producing the maximal number of offspring. It thus seems to result from the 'inertia' (i.e., the over-generalization) of an adaptive mechanism that has established itself in the brain during evolution. Incorrectly identifying mental phenomena adaptations as malfunctions instead of adaptations may have severe consequences. Nesse describes that depression can be the manifestation of a malfunction but may also, on the other hand, just as well be the result of a natural defence, or even a natural defence gone awry.[21] For comparison, bodily ailments can arise directly from defects like jaundice or seizures, or defence mechanisms like pain or diarrhoea. For clinical purposes, the distinction is vital. Curing a cough starts with searching for the abnormalities that lead to it, not the suppression of the cough. Similarly, treating a depression with psychotropic drugs without searching for the cause may be detrimental in the long run. In general, to advance our understanding of the ultimate causes of behaviour requires all these options to be taken into account.

2. Modelling Methods for the Study of Adaptive Behaviour

Instead of trying to reconstruct evolutionary development by collecting empirical data, it may be more productive to use a modelling approach. Biologists construct mathematical evolutionary models that

[21] R.M. Nesse, *Is Depression an Adaptation?*, "Archives of General Psychiatry" 2000; vol. 57, pp. 14–20.

are typically used to calculate gene frequency in relation to a set of parameters. If the genes code for behaviour, this sheds light onto the theoretical adaptive value of such behaviour. For instance, Beltman and Haccou studied speciation in relation to a hypothetical organism's ability to learn habitat features.[22] In their model, this trait led to the development of a habitat preference, which reinforced habitat adaptation, and ultimately led to speciation. Insightful conclusions can be drawn from models such as these. However, incorporating extra complexity, such as neural energy consumption, easily renders the mathematics of the model intractable. Simulation models are not limited in this way and can be extended to represent such interactions.

2.1 Genetic Algorithms and Evolutionary Simulations

A very fruitful approach which mimics evolution with computer simulations, allows for the integration of behavioural and neural aspects in the models. Moreover, this approach may help avoid many of the previously discussed problems with evolutionary psychology. Evolutionary computation is mostly performed with Genetic Algorithms.[23] Goldberg defined Genetic Algorithms as "... search algorithms based on the mechanics of natural selection and natural genetics".[24] Like DNA in biological systems, Genetic Algorithms encode the individual 'solutions' to a problem in a coded parameter set (usually with binary features). The search for optimal solutions starts with an initial population that consists of a random combination of parameter values and proceeds through reproduction. Fitness is defined as some measure of how well the parameter set solves the problem. Thus, better solutions

---

[22] J.B. Beltman, P. Haccou, *Speciation Through the Learning of Habitat Features*, "Theoretical Population Biology" 2005, vol. 67, pp. 189–202.

[23] J.H. Holland, *Adaptation in Natural and Artificial Systems*, University of Michigan Press, Ann Arbor 1975.

[24] D.E. Goldberg, *Genetic Algorithms in Search, Optimization and Machine Learning*, Addison-Wesley, Reading, Mass. 1989.

get a higher chance of survival and reproduction. When parameter sets are reproduced, they are subject to mutations and crossovers similar to natural reproduction. Mutation may improve the performance of the population by occasionally suggesting a new partial solution. The mutation probability of each bit is usually low. With crossover, the parts of two individual solutions are recombined into two, possibly very different, solutions. The newly combined solutions replace their less fit counterparts. From generation to generation, this leads to a higher overall fitness.

By applying a Genetic Algorithm to artificial neural networks that control the behaviour of agents in a virtual environment, adaptations to the specific environment can develop. Over generations, the individuals with networks that produce behaviour yielding the highest odds for reproduction and survival propagate their genetic information furthest. In general, through simulations it is possible to perform experiments that would otherwise be too costly, too time consuming, unethical or even impossible. In contrast to real-life experiments, the programmer has complete power to manipulate the self-created world (i.e., the initial conditions), but has to wait and see what develops. Only limitations of computer power and time apply.[25]

The technique has become increasingly popular in disciplines ranging from chemistry and physics to economics and linguistics. In biology, an impressive example is provided by a simulation that reached the ears of the world press in 1994, performed by the zoologists Dan-Erik Nilsson and Susanne Pelger.[26] They countered the famous remark by Bishop Paley that the human eye surely would have to have been designed, something which has aroused debate ever since Darwin first published the Origin of Species in 1859. Nilsson and Pelger performed simulations to make an estimation of the time required for an eye to evolve by gradual changes and natural se-

---

[25] S.L. Peck, *Simulation as Experiment: A Philosophical Reassessment For Biological Modelling*, "Trends in Ecology and Evolution" 2004, vol. 19, pp. 530–534.

[26] D-E. Nilsson, S. Pelger, *A Pessimistic Estimate of the Time Required For an Eye to Evolve*, "Proceedings of the Royal Society London B" 1994, vol. 256, pp. 53–58.

lection. They started out with a light sensitive patch that could only mutate minutely and after many generations this eventually led to a full-fledged eye. Their work reached the world press, not because it showed how an eye evolved, but because it showed that also an eye could gradually be constructed by evolution. Using minimal assumptions for their calculations, they even concluded that the transformation of a light sensitive patch to a focused lens eye would have only required a few hundred thousand years. Whether these computational results sufficiently resemble biological evolution, of course, remains a matter of debate. Whatever the outcome, simulations remain a highly practical way to perform experiments on evolutionary development.

Classical evolutionary simulation studies[27] are simply implementations of the mathematical models that calculate gene frequency.[28] The computational method has the advantage that it can easily be extended to allow a greater level of complexity than methods that only study evolution analytically. In its simplest form, similar to the mathematical models, the genes in the Genetic Algorithms map directly to a behaviour, strategy or preference. Reggia, Schulz, Wilkinson and Uriagereka, for instance, studied the adaptive value of signalling others when an agent detects food or a predator.[29] In their model, two genes that could either be on or off determined the traits whether an agent alerted those around it in either situation or not. They found that the trait to alert others to the presence of a predator evolved easily. The gene occurred in high frequencies after only few generations and under a large set of parameter variations. Alerting for food depended heavily on food density, the number of predators and the fierceness of competition. In addition, the priority of food over longevity, which

[27] R.J. Collins,D.R. Jefferson, *AntFarm: Towards Simulated Evolution,* [in:] *Artifical Life II,* eds. C.G. Langton, C. Taylor, J.D. Farmer, S. Rasmussen, Addison-Wesley, Redwood City, CA 1992, pp. 579–601.

[28] J.B. Beltman, P. Haccou, *op. cit.*

[29] J.A. Reggia, R. Schulz, G. Wilkinson, J. Uriagereka, *Conditions Enabling the Evolution of Inter-Agent Signaling in an Artificial World,* "Artificial Life" 2001, vol. 7, pp. 3–32.

depended on the chosen fitness function, was of great influence for the development of the food signalling trait. Because agents that alert others when they detect a predator spend more time fleeing, the overall food consumption of these agents would be lower. Hence, increasing the odds for reproduction dependent on food consumption could actually decrease the trait frequency for predator signalling.

These findings can be matched to the behaviour of several species and, although the results are thus interesting in their own right, the model never considers the neural interaction between the two behaviours. It would be reasonable to suppose that the underlying neural substrate has at least some overlap and genetic modification of the one trait could very well directly influence the other. Taking this consideration into account may provide more realistic insights into the evolution of signalling others for food and predators.

In economics, Axelrod started a productive line of research in 1980 using agents in an evolutionary process. He studied negotiation and cooperation strategies between agents, and groups of agents, measuring their success in playing a prisoner's dilemma game or other types of ultimatum games.[30] In these simulations, genes typically coded directly for negotiation strategies. Of all those that were tested against each other the 'tit for tat' strategy (i.e., always initially cooperating and retaliating only when provoked) proved highly successful under many circumstances. Unfortunately, the gene for behaviour shortcut, or gene for mental function shortcut made in these simulations, disconnects the conclusions from the rest of psychology and cognitive neuroscience. In this example, the practical application of this discovery from game theory could benefit tremendously from insights into the proximate mechanisms at play during negotiations.

The limitation of treating behaviour, such as a negotiating strategy, as a particular fundamental trait, is effectively demonstrated by an experiment described by Brosnan and De Waal with capuchin

---

[30] R. Axelrod, W.D. Hamilton, *The Evolution of Cooperation*, "Science" 1981, vol. 211, pp. 1390–96.

monkeys.[31] They report that in the first phase of the experiment the capuchin monkeys were willing to exchange a token (a small granite rock) for a reward (a piece of cucumber). Hence, although perhaps not consciously, the monkey's negotiating strategy dictated that the trade, rock for cucumber, was acceptable. In the second phase, however, the monkey could observe its neighbour receiving grapes (which were deemed much tastier) for the same rock. In this case, in over half of the trials, the monkey would simply refuse the cucumber, and on occasion show fits of anger, throwing the reward violently back at the human experimenter. The interpretation of these results as an aversion of inequity has been challenged by a follow-up study, in which cucumbers were equally refused if only the grape was present in the adjacent cage.[32] Perhaps it is not the inequity of pay and just the fact that a better reward is present that drives the refusal. Although elaborate cognitive deliberation was first invoked on the part of the monkey, more basic evaluation processes that have been extensively researched in traditional psychology may be responsible for this behaviour.

Even so, irrespective of the outcome of this debate, both experiments clearly demonstrate that there are more psychological functions involved in a monkey's mind when bartering than bare cost-benefits analyses. Reducing the complex minds of humans to such terms seems inapt even more so. The experiment exemplifies the notion that the reduction a behaviour's cause to a single inheritable trait ignores the underlying complex interactions and influences across mental functions and neural structures. Regrettably, the value of such 'genes for behaviour' studies, which leave the gap between the proximate and ultimate mechanisms unabridged, remain limited. These restrictions may be overcome by simulating the evolution of more complex models incorporating both behavioural and neural features.

---

[31] S.F. Brosnan, F.B.M. De Waal, *Monkeys Reject Unequal Pay*, "Nature" 2003, vol. 425, pp. 297–299.

[32] C.D.L.Wynne, *Animal Behaviour: Fair Refusal by Capuchin Monkeys*, "Nature" 2004, vol. 428, p. 140.

2.2 Synthetic Evolutionary Psychology

Traditionally, models are constructed to comprehend and possibly predict empirical data – for instance, models constructed with formalisms as ACT-R by Anderson[33] or Soar by Laird, Newell & Rosenbloom.[34] Conversely, models emerging from evolutionary computation (i.e., evolutionary growth) may precede behavioural analysis. It can be seen as a kind of synthetic psychology,[35] which starts out with the evolution of basic mechanisms and only later investigates the full range of behavioural consequences. Extending Braitenberg's[36] concept of synthetic psychology, Evans and de Back have coined the term 'synthetic evolutionary psychology' for the approach that applies evolutionary simulations to psychology.[37] They give the following description: "Synthetic methods involve building models of the system under investigation and then observing their behaviour. The more closely the behaviour of the model corresponds to the behaviour of the target system, the more confident we can be that the internal structure of the model corresponds to the internal structure of the target".[38] The synthetic approach thus provides evolutionary psychologists with the tool they lack for the study of the ultimate causes of behaviour, and lead to more attention in evolutionary quarters to the proximate causes of behaviour, on which a large body of knowledge has been gathered in classical psychology. By simulating the evolutionary de-

[33] J.R. Anderson, *The Architecture of Cognition,* Harvard University Press, Cambridge, Mass. 1983.

[34] J.E. Laird, A. Newell, P.S. Rosenbloom, *Soar: An Architecture for General Intelligence*, "Artificial Intelligence" 1987, vol. 33, pp. 1–64.

[35] V. Braitenberg, *Vehicles: Experiments in Synthetic Psychology*, MIT Press, Cambridge, Mass. 1984; cf. M.R.W. Dawson, *From Embodied Cognitive Science to Synthetic Psychology*, [in:] *Proceedings of the First IEEE International Conference on Cognitive Informatics (ICCI'02)*, eds. Y. Wang, R.H. Johnston, M.R. Smith, IEEE Computer Society, Los Alamitos, CA 2002, pp. 13–22.

[36] V. Braitenberg, *op. cit.*

[37] D. Evans, W. de Back, *Evolutionary Robotics*, [in:] *Applied Evolutionary Psychology*, ed. C. Roberts, Oxford University Press, Oxford 2012, pp. 414–425.

[38] *Ibid.*, p. 419.

velopment of basic mechanisms it is possible to investigate under which environmental conditions such a particular basic mechanism is most likely to develop. If mechanisms can emerge in both simple and complex environments, with few or many constraints, then principles of parsimony dictate that the simpler conditions are more likely to have played a plausible role in biological evolutionary history. When considering competing models that fit the available data equally well, the model that is more likely to develop in evolutionary simulations is preferred above the others. It is important to note that with this approach the modeller does not construct the model, but merely sets up initial conditions that are conducive to the gradual development of the model that performs a particular function. Due to the autonomic synthetic quality of the simulations, the emerging networks can lead to new mechanisms and hypotheses, not previously considered by psychology.[39]

2.3 Synthetic Evolutionary Cognitive Neuroscience

Ideally, the proximal causes and the basic mechanisms should be formulated in the neural language that is commonly used in neural network models. The inclusion of neural functions and mechanisms allows for the incorporation of knowledge from cognitive neuroscience.[40] The neural costs of these mechanisms may, moreover, explain behaviour without obvious benefits to the organism. Using a more sophisticated approach, the genes in Genetic Algorithms may code for the configuration of neural structures. In such evolutionary simulations, neurons are joined together in a network that produces output and so determine the behaviour of a particular agent. In turn, the behaviour leads to a change in input to the receptors of the network.

---

39 B.T. Heerebout, R.H. Phaf, *Emergent Oscillations in Evolutionary Simulations*, "Journal of Cognitive Neuroscience" 2010, vol. 22, pp. 807–823.

40 S. Haykin, *Neural Networks: A Comprehensive Foundation*, Prentice-Hall International, London 1999.

Eventually, the aggregate behaviour of the agent influences its odds for survival and reproduction. Synthetic evolutionary cognitive neuroscience may thus weigh behavioural benefits and neural costs in a more biologically plausible manner.

Mimicking the evolutionary process to construct artificial neural networks has been gaining in popularity. Batali and Grundy, for instance, simulated the evolution of artificial neural networks that functioned as a 'motivational system'.[41] In their simulations, a 'reinforcement' network with hereditary weights evaluated the output of a learning 'action' network. The 'reinforcement' networks could evolve to reward the beneficial output of the 'action' networks, and so directing the learned behaviour. However, the 'physical rules' that determined whether an action was beneficial consisted of simple binary mappings. The networks did not simulate 'embodied' free moving agents. Other roboticists, on the other hand, have applied the evolutionary approach to robots by growing artificial neural networks that could perform intelligent tasks.[42] Using evolutionary simulations with embodied agents to investigate the adaptive value of specific models from cognitive neuroscience remain rare, with the exception of work performed by John Bullinaria[43] – for an even more recent one, see Ward and Ward.[44] He took Genetic Al-

[41] J. Batali, W.N. Grundy, *Modeling the Evolution of Motivation*, "Evolutionary Computation" 1996, vol. 4(3), pp. 235–270.

[42] E.g. F. Gruau, *Neural Network Synthesis Using Cellular Encoding and the Genetic Algorithm*, Doctoral dissertation, Ecole Normale Superieure de Lyon, France 1994; K.O. Stanley, R. Miikkulainen, *Achieving High-level Functionality through Complexification*, [in:] *Proceedings of the AAAI-2003 Spring Symposium on Computational Synthesis*, 2003, pp. 226–232; F. Dellaert, R.D. Beer, *Toward an Evolvable Model of Development for Autonomous Agent Synthesis*, [in:] *Proceedings of Artificial Life IV*, eds. R. Brooks, P. Maes, MIT Press, Cambridge, Mass. 1994, pp. 246–257.

[43] J.A. Bullinaria, *Simulating the Evolution of Modular Neural Systems*, [in:] *Proceedings of the Twenty-third Annual Conference of the Cognitive Science Society*, Lawrence Erlbaum Associates, Mahwah, NJ 2001, pp.146–151, see also B.L.M. Happel, J.M.J. Murre, *Design and Evolution of Modular Neural Network Architectures*, "Neural networks" 1994, vol. 7, pp. 985–1004.

[44] R. Ward, R. Ward, *Selective Attention and Control of Action: Comparative Psychology of an Artificial, Evolved Agent and People*, "Journal of Experimental Psychology: Human Perception and Performance" 2008, vol. 34, pp. 1165–1182.

gorithms beyond the use as an optimization tool and applied the technique to study the adaptive value of modularity in these networks.

Modularity is a well-studied phenomenon in cognitive neuroscience. The advantages of modularity, such as the brain's separate 'where' and 'what' vision pathways, have been demonstrated by connectionists Rueckl, Cave, and Kosslyn.[45] Their work was complemented by Jacobs, Jordan and Barto showing that modularity can develop if a gating network filters the error feedback only to the best performer of a set of competing networks.[46] For different tasks different networks will win the competition and because only the winner receives feedback to learn from, task dependent expertise develops in the networks. Bullinaria investigated whether this form of modularity could evolve in an evolutionary process.[47] He performed simulations in which a population of parameter sets evolved. The parameter set defined the configuration of the neural networks. In his rather concise conclusion he stated that: "...the evolution of modularity depends crucially on the choice between competing non-biologically plausible learning algorithms. Whilst it is easy to find reasons why modular neural systems should be superior to their non-modular counterparts, we have seen that simulating the evolution of such modularity is not as straightforward as it is often assumed".[48] If mimicking evolution is less straightforward than often assumed, this is probably because evolution itself is an onerous concept and should only consider biologically plausible neural costs.

---

[45] J.G. Rueckl, K.R. Cave, S.M. Kosslyn, *Why are "What" and "Where" Processed by Separate Cortical Visual Systems? A Computational Investigation*, "Journal of Cognitive Neuroscience" 1989, vol. 1, pp. 171–186.

[46] R.A. Jacobs, M.I. Jordan, A.G. Barto, *Task Decomposition Through Competition in Modular Connectionist Architecture: The What and Where Vision Tasks*, "Cognitive Science" 1991, vol. 15, pp. 219–250; for another conception of modularity see J.M.J. Murre, R.H. Phaf, G. Wolters, *CALM: Categorizing and Learning Module*, "Neural Networks" 1992, vol. 5, pp. 55–82.

[47] J.A. Bullinaria, *op. cit.*

[48] *Ibid.*, p. 151.

As another example, the roboticists Nolfi and Parisi have applied evolutionary simulations to study adaptive intelligent behaviour.[49] They created an environment with embodied agents that had to search for food to produce offspring. The initial weights of the neural networks were determined by their 'genes'. However, during their lifetime the weights also changed as they learned to predict what the next sensor input would be as a function of their current input and movement. The learned ability to predict the next input would not necessarily improve an agent's fitness. Moreover, the learned ability itself was, of course, not passed on to the next generation and it did thus not influence evolution directly. However, for some agents the predicting ability did help in their effort to collect food. Those agents could reach higher fitness levels before others and so reproduced more often. In the last generations the agents' learning skill had improved significantly compared to that of the initial agents. This is remarkable, because the evolutionary process did not favour the predicting ability by itself. Moreover, the learning skill could only be modified by changing the same weights that were also responsible for the food collection. Because the agent's initial performance in predicting the subsequent input was constant over all generations, it could only be concluded that the evolutionary process prepared the agents to learn through the mutation of the genes encoding these weights.

The results emerging from the simulations by Nolfi and Parisi[50] come very close to the hypothesis of Öhman and Mineka[51] on the evolutionarily preparation of the mammalian brain for learning to recognize and react to the danger of specific stimuli. This evolutionary

---

[49] S. Nolfi, D. Parisi, *Neural Networks in an Artificial Life Perspective*, [in:] *Artificial Neural Networks (ICANN97). Proceedings of the 7th International Conference on Artificial Neural Networks*, eds. W. Gerstner, A. Germond, M. Hasler, J.D. Nicoud, Springer Verlag, Berlin 1997, pp. 733–738; S. Nolfi, D. Parisi, *Evolution of Artificial Neural Networks*, [in:] *Handbook of Brain Theory and Neural Networks*, ed. M.A. Arbib, MIT Press, Cambridge, Mass. 2002, pp. 418–421.

[50] *Ibid.*

[51] A. Öhman, S. Mineka, *Fears, Phobias, and Preparedness: Toward an Evolved Module of Fear and Fear Learning*, „Psychological Review" 2001, vol. 108, pp. 483–522.

preparation does not entail that these reactions are innate or fully determined by evolution, but imply that they are merely privileged in subsequent learning processes. Öhman and Mineka described experiments where they showed short video clips to rhesus monkeys which had been born in captivity.[52] The clips depicted other monkeys reacting anxiously to either a toy snake or a colourful flower. Although the stimuli were matched for saliency and other confounding variables, after only a few trials the subject monkeys also displayed anxiety when confronted with the toy snake, but not when confronted with the flower. Öhman and Mineka explained this by assuming that evolution affects the same connections in the neural mechanism as learning. Evolution has given the connection strengths associated with the evolutionary prepared stimuli a slight advantage over connections associated with other stimuli, which biases competition towards these privileged stimuli in the learning process.

3. The Evolution of Affective Processing

Emotional systems are prime examples of modes of behaviour that are shared by all members of the human species, and possibly by other species too.[53] Similar neural architectures associated with these emotional modes, have been identified in all mammals and at least one species of bird (i.e., pigeons).[54] They are presumed to have evolved already early in evolution.[55] The analysis of emotions should therefore go beyond behavioural observations and an intuitive understanding in

[52] Cf. S. Mineka, M. Cook, *Immunization Against the Observational Conditioning of Snake Fear in Rhesus Monkeys*, "Journal of Abnormal Psychology" 1986, vol. 95, pp. 307–318

[53] E.g. K. Oatley, P.N. Johnson-Laird, *Towards a Cognitive Theory of Emotions*, "Cognition and Emotion" 1987, vol. 1, pp. 29–50.

[54] E.g. J.E. LeDoux, *Sensory Systems and Emotion: A Model of Affective Processing*, "Integrative Psychiatry" 1986, vol. 4, pp. 237–248; J.E. LeDoux, *The Emotional Brain*, Simon & Schuster, New York 1996.

[55] J. Panksepp, J.B. Panksepp, *The Seven Sins of Evolutionary Psychology*, "Evolution and Cognition" 2000, vol. 6(2), pp. 108–131.

terms of personal experience. An evolutionary grounding of affective value has been suggested by Johnston.[56] He proposed that positive affect constitutes the neural code of those conditions that enhance evolutionary fitness, whereas negative affect codes for fitness-reducing circumstances. Viewing emotions thus as motivational states, producing evolutionary adaptive responses to stimuli, conveniently allows for comparison with the evolved behaviour of reproducing, minimally cognitive agents.[57] The dynamic simulations of embodied agents from our group may serve as a pioneering example of a synthetic evolutionary approach to emotions, and to fear in particular.[58]

3.1 Dual-Pathway Architectures

Den Dulk and colleagues employed the simulation approach to investigate evolutionary hypotheses about emotional processes and their neural substrate.[59] LeDoux's dual-pathway model suggested that sensory input related to fear stimuli can reach the amygdala via two parallel pathways: one comes directly from the sensory thalamus and the other loops through the cortex.[60] Processing in the direct path is faster, has less capacity, and is more coarsely grained than the indirect pathway. The amygdala, which is located deep in the medial

---

[56] V.S. Johnston, *The Origin and Function of Pleasure*, "Cognition and Emotion" 2003, vol. 17, pp. 167–179; for a further elaboration of these ideas see also R.H. Phaf, M. Rotteveel, *Affective Monitoring: A Generic Mechanism For Affect Elicitation*, "Frontiers in Psychology" 2012, vol. 3, p. 47.

[57] R.D. Beer, *Toward the Evolution of Dynamical Neural Networks for Minimally Cognitive Behaviour*, [in:] *From Animals to Animats 4: Proceedings of the Fourth International Conference on Simulation of Adaptive Behavior*, eds. P. Maes, M. Mataric, J. Meyer, J. Pollack, S. Wilson, MIT Press, Cambridge, Mass. 1996, pp. 421–429.

[58] P. den Dulk, B.T. Heerebout, R.H. Phaf, *A Computational Study into the Evolution of Dual-route Dynamics for Affective Processing*, "Journal of Cognitive Neuroscience" 2003, vol. 15, pp. 194–208.

[59] *Ibid.*

[60] J.E. LeDoux, *Sensory Systems and Emotion: A Model of Affective Processing*, "Integrative Psychiatry" 1986, vol. 4, pp. 237–248; J.E. LeDoux, *The Emotional Brain*, Simon & Schuster, New York 1996.

temporal lobes, presumably plays a primary role in the elicitation of fear and the initiation of fear responses. Because the direct pathway to the amygdala has a lesser ability to differentiate between stimuli, a fear response may be evoked unnecessarily (i.e. a "false alarm"). The short pathway to the amygdala is able to meet the urgency of the situation, but is also liable to produce such false alarms. When there is more time to process stimuli extensively, the costs of a false alarm may even be reduced by the longer but slower pathway through inhibiting the direct action tendencies. LeDoux explained the dual-pathway architecture by arguing that in threatening situations the evolutionary costs of a 'miss' (react too late) heavily outweigh the cumulative costs of all false alarms. The very innovative aspect of LeDoux's reasoning is that it directly concerned the evolution of a neural architecture. His argument focused on the internal cognitive, and affective, processes instead of on externally observable behaviour, as was the case in the older ethological approaches.[61]

For the design of the simulations, den Dulk et al. partly copied Beer's simulation method.[62] He investigated the dynamics of evolving agents, controlled by artificial neural networks, in a virtual environment. His agents had to navigate a continuous world in which they searched plants for food using chemotaxis, which means that they had to home in on the smells emitted by the plants. The weights of the network connections constituted the genes of the agent. Den Dulk et al. added biologically plausible processing delays, which consisted of time costs for updating the nodes' activations. Instantaneous processing of stimuli by networks is simply unrealistic, because synaptic transfer in the nervous system always takes time. Time, moreover, plays a crucial role as a neural cost in LeDoux's evolutionary reasoning.

---

[61] C. Darwin, *The Expression of the Emotions in Man and Animals*, University of Chicago Press, Chicago 1872/1965; K. Lorenz, *Evolution of Ritualisation in the Biological and Cultural Spheres*, "Philosophical Transactions of the Royal Society of London" 1966, vol. 251, pp. 273–284.

[62] R.D. Beer, *Intelligence as Adaptive Behavior: An Experiment in Computational Neuroethology*, Academic Press, San Diego, CA 1990.

Den Dulk et al. also added potentially fitness-reducing predators that the agents had to avoid while searching the plants. An agent gained energy by collecting food until it was touched (i.e., 'eaten') by a predator. In addition, energy depletion increased with speed of movement. The neural network configuration of an agent's offspring resembled that of the parent's with slight variations in the strength of the network connections through mutation and cross-overs. The agents' reproductive success depended on their fitness level, which combined survival length and the amount of energy gathered in the agent's lifetime.

After one thousand generations of simulated evolution, agents with dual-processing networks resembling LeDoux's architecture developed. Processing in the direct pathway of these agents induced the fast avoidance of both predator and food, whereas slower indirect processing led to food approach and facilitated predator avoidance. These agents were only found in simulations when food and predator stimuli, (i.e., the smells they emitted), were hard to distinguish and the fitness reflected time pressures in escaping from the predator. These outcomes illustrate our earlier observation that a single behaviour may seem maladaptive (e.g., the initial avoidance of food), but that it makes perfect evolutionary sense when also the neural substrate is taken into account. Such seemingly adverse behaviour can, nonetheless, emerge in evolution because the balance of the behavioural benefits and neural costs of the underlying neural structure is very much positive. At the core of LeDoux's dual pathway hypothesis is the trade-off between speed and accuracy, born from the constraints imposed on the neuronal processing by the laws of physics. Without a time cost of synaptic transfer, the dual-processing architecture would not have developed.

Undoubtedly, the actual fear network is much more complicated than the artificial neural network described here and the neural localization of these structures cannot be definitively checked by these simulations. Nevertheless, the converging neurobiological and computational results provide strong support for LeDoux's general architecture with parallel processing pathways balancing time costs and complexity.

3.2 Emerging Oscillations Improve Switching Ability

A newer set of simulations from our group demonstrate the creative power of the synthetic approach.[63] In these simulations, the networks could develop a recurrent structure (see Figure 1). In the new networks, the hidden layer projected to the output layer, and to an additional layer with recurrent projections to the sending hidden layer. As a result, a negative feedback mechanism could evolve that produced oscillatory activation patterns.[64] Surprisingly these oscillations greatly improved the agents' fitness, which reflects on their ability to collect food and avoid predators.

Further analyses, in which the agents first responded to a plant that would suddenly be replaced by a predator, revealed that the oscillations specifically enhanced the behavioural switching efficacy. In comparison, the networks of agents that evolved in the control simulation, without the basic recurrent structure, could distinguish plants from predator with similar precision, but reactions were slower to turn and adjust speed. The enhanced behavioural switching ability, moreover, allowed for a greater range of behavioural differentiation in response to plants and predators. A distinction between attentional selection and the modulation of attention emerged in these simulations. The nature of the selected action was not changed and only the switching between actions was modulated by the oscillations. A new type of dual-pathway architecture also arose. The indirect route was now solely responsible for the selection of specific action tendencies, but a fast 'energizing' by the direct route was required for the transformation of the selected action tendency into action.

---

[63] B.T. Heerebout, R.H. Phaf, *Emergent Oscillations in Evolutionary Simulations*, "Journal of Cognitive Neuroscience" 2010, vol. 22, pp. 807–823.

[64] Cf. R. Ritz, T.J. Sejnowski, *Synchronous Oscillatory Activity in Sensory Systems; New Vistas on Mechanisms*, "Current Opinion in Neurobiology" 1997, vol. 7, pp. 536–546; D. Dupret, B. Pleydell-Bouverie, J. Csicsvari, *Inhibitory Interneurons and Network Oscillations*, "Proceedings of the National Academy of Sciences" 2008, vol. 105, pp. 18079–18080.

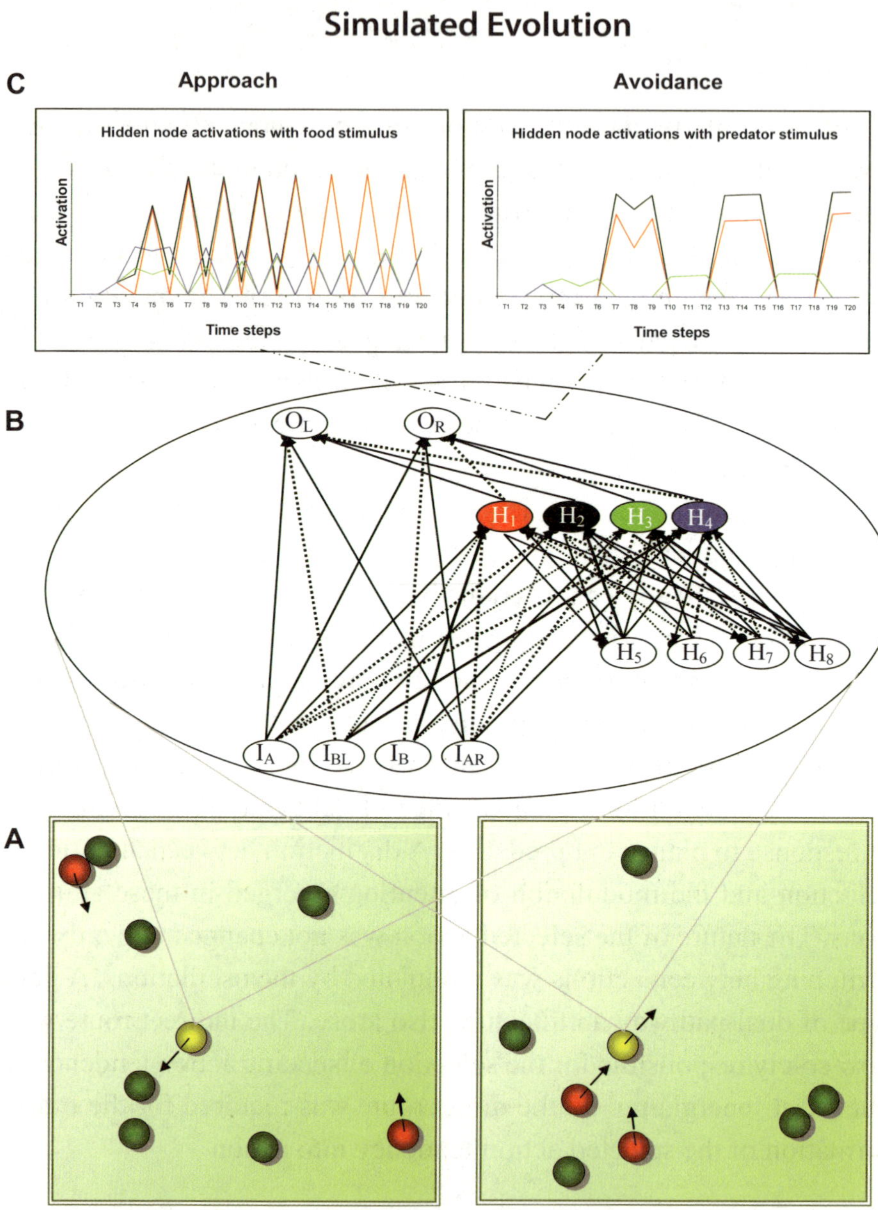

Figure 1. Graphical display of agents with oscillating networks. (A) Two virtual environments with an agent (yellow), plants (green) and predators (red). In the environment on the left the agent senses the plants and moves in their direction. On the right, the agent flees from two hunting predators. (B) Schematic view of the neural network governing the behaviour of the agents. (C) The activation values of the nodes of the network's hidden layer plotted over time. The graphs on the left, showing fast oscillations, are of the agent responding to the plants. On the right, the nodes' activations oscillate slow as the agent flees from the predator.

A simple model, distilled from the evolved networks, clarified the increased switching efficacy resulting from the oscillations. The advantage could be explained as a beneficial addition to the competitive process of attentional selection.[65] The most biologically plausible implementation of competition in a neural network is by mutual inhibition, which results in a type of contrast enhancement. This is a basic neural mechanism, which not only emerged in our simulations but also seems prevalent in the brain.[66] If a winner continues to inhibit all its competitors after the resolution of competition, it can impede the transition to other, appropriate behaviour when a new external situation arises. However, if all nodes, including the winners, are periodically inhibited, competition is refreshed and new competitors get a chance to take over. In this manner the oscillations operate to facilitate a switch of winners in the network.

3.3 Oscillation Frequency and Affect

Following the rationale that oscillations facilitate switching, it would make sense if the frequencies of the oscillations differed per stimulus type. After all, a switch in response is required with great urgency when a foraging agent suddenly detects a predator. In contrast, if the agent is fleeing from a predator, it is best not to be distracted by some food patch. Therefore, higher oscillation frequencies, enabling faster switches, were expected when an agent responded to food than when it responded to a predator. Analysis of all agents from each last

---

[65] Cf. J. Duncan, *Cooperating Brain Systems in Selective Perception and Action*, [in:] *Attention and Performance, Vol. XVI. Information Integration in Perception and Communication*, eds. T. Inui, J.L. McClelland, MIT Press, Cambridge, Mass. 1996, pp. 549–578; R.H. Phaf, A.H.C. Van der Heijden, P.T.W. Hudson, *SLAM: A Connectionist Model for Attention in Visual Selection Tasks*, "Cognitive Psychology" 1990, vol. 22, pp. 273–341.

[66] P. Fries, D. Nikolic, W. Singer, *The Gamma Cycle*, "Trends in Neuroscience" 2007, vol. 30, pp. 309–316; C. Börgers, S. Epstein, N.J. Kopell, *Gamma Oscillations Mediate Stimulus Competition and Attentional Selection in a Cortical Network Model*, "Proceedings of the National Academy of Sciences" 2008, vol. 105, pp. 18023–18028.

generation showed that the networks indeed oscillated in a shorter cycle with plant stimuli than with predator stimuli.[67] Figure 1 shows an agent in different situations, its network structure and its responding oscillatory patterns. The recorded difference in oscillation frequencies also adds further support to the initial conclusion that the adaptive value of the oscillations was caused by an increased switching efficacy of the networks.[68]

The agents in the simulations switched their behaviour faster while shifting from responding to 'positive' stimuli to 'negative' than the other way around. This agrees very well with the ample empirical evidence that positive affect increases cognitive flexibility.[69] In particular, the task switching paradigm by Tan, Jones, and Watson[70] demonstrated that positive mood induction leads to increased attentional flexibility in human participants.[71] Although the influence of affect on attentional flexibility has been well studied and several recent stud-

---

[67] B.T. Heerebout, R.H. Phaf, *Good Vibrations Switch Attention: An Affective Function for Network Oscillations in Evolutionary Simulations*, "Cognitive, Affective, & Behavioral Neuroscience" 2010, vol. 10, pp. 217–229.

[68] See B.T. Heerebout, R.H. Phaf, *Emergent Oscillations in Evolutionary Simulations*, "Journal of Cognitive Neuroscience" 2010, vol. 22, pp. 807–823.

[69] N. Baumann, J. Kuhl, *Positive Affect and Flexibility: Overcoming the Precedence of Global over Local Processing of Visual Information*, "Motivation and Emotion" 2005, no 29, pp. 123–134; E. Das, B.M. Fennis, *In the Mood to Face the Facts: When a Positive Mood Promotes Systematic Processing of Self-threatening Information*, "Motivation and Emotion" 2008, vol. 32, pp. 221–230; G. Dreisbach, T. Goschke, *How Positive Affect Modulates Cognitive Control: Reduced Perseveration at the Cost of Increased Distractibility*, "Journal of Experimental Psychology" 2004, vol. 30, pp. 343–353; M.J. Fenske, J.D. Eastwood, *Modulation of Focused Attention by Faces Expressing Emotion: Evidence From Flanker Tasks*, "Emotion" 2003, vol. 3(4), pp. 327–343; B.L. Fredrickson, *The Broaden-and-build Theory of Positive Emotions*, "Philosophical Transactions of the Royal Society B: Biological Sciences" 2004, vol. 359, pp. 1367–1377; A.M. Isen, *Positive Affect*, [in:] *Handbook of Cognition and Emotion*, eds. T. Dalgleish, M. J. Power, Wiley & Sons, New York 1999, pp. 521–539; P. Luu, D.M. Tucker, D. Derryberry, *Anxiety and the Motivational Basis of Working Memory*, "Cognitive Therapy and Research" 1998, vol. 22, pp. 577–594.

[70] H.K. Tan, G.V. Jones, D.G. Watson, *Encouraging the Perceptual Underdog: Positive Affective Priming of Nonprefererred Local-global Processes*, "Emotion" 2009, vol. 9, pp. 238–247.

[71] See also B.T. Heerebout, A. Todorović, H.E. Smedinga, R.H. Phaf, *Affective Modulation of Attentional Switching*, "American Journal of Psychology", (in press).

ies correlate high frequency oscillations with attention, to our knowledge, there is almost no empirical research that explicitly investigates the relationship between neural oscillations and affect.[72] Based on the results from the evolutionary simulations, we would predict that high neural oscillation frequencies (i.e., particularly in the gamma band 30–80 Hz) enable efficient attentional switching associated with positive affective states, whereas low frequencies foster a locking of attentional focus to negative stimuli.

The imaginative study by Tsai and collaborators, may be one of the few to support a causal relationship between gamma oscillations and positive affect.[73] These researchers found that only high frequency, but not lower frequency, oscillations induced in dopaminergic neurons could substitute the reward signal in a place preference task. Inspired by this study, we also made a first attempt to study the relation between high frequency oscillations and affect empirically.[74] We adapted an attention-shift paradigm by Bauer, Cheadle, Parton, Müller, and Usher.[75] These researchers induced gamma oscillations by presenting visual gamma flicker on a screen.[76] The low contrast gamma flicker, which the participants were not aware of, served as a prime on one of three locations where the targets appeared. Bauer et al.'s results indeed indicated an association between gamma

---

[72] E.g. J. Fell, G. Fernández, P. Klaver, C.E. Elger, P. Fries, *Is Synchronized Neuronal Gamma Activity Relevant for Selective Attention?*, "Brain Research Reviews" 2003, vol. 42, pp. 265–272.

[73] H. Tsai, F. Zhang, A. Adamantidis, G.D. Stuber, A. Bonci, L. de Lecea, K. Deisseroth, *Phasic Firing in Dopaminergic Neurons is Sufficient for Behavioral Conditioning*, "Science" 2009, vol. 324, pp. 1080–1084.

[74] B.T. Heerebout, A.E.Y. Tap, M. Rotteveel, R.H. Phaf, *Gamma Flicker Elicits Positive Affect Without Awareness*, (submitted).

[75] F. Bauer, S.W. Cheadle, A. Parton, H.J. Müller, M. Usher, *Gamma Flicker Triggers Attentional Selection without Awareness*, "Proceedings of the National Academy of Sciences" 2009, vol. 106, pp. 1666–1671.

[76] See also C.S. Herrmann, *Human EEG Responses to 1–100 Hz Flicker: Resonance Phenomena in Visual Cortex and Their Potential Correlation to Cognitive Phenomena*, "Experimental Brain Research" 2001, vol. 137, pp. 346–353; P.E. Williams, F. Mechler, J. Gordon, R. Shapley, M.J. Hawken, *Entrainment to Video Displays in Primary Visual Cortex of Macaque and Humans*, "Journal of Neuroscience" 2004, vol. 24, pp. 8278–8288.

oscillations and attentional shifting. In our adaptation, the participants had to select a target of deviating sex from three neutral faces, either one female (target) and two males (distracters) or one male (target) and two females (distracters). The participants had to rate the target face as either pleasant or unpleasant. The main finding was that gamma flicker led to positive affect, both when measured implicitly and explicitly. The explicit rating of neutral faces, for instance, went up from 48% in the no-flicker condition to 55% in the 50 Hz flicker condition.

These results are still tentative and further experiments with human participants are certainly required. Moreover, the neurophysiological base for the mechanisms asserted here is still very much underspecified. But regardless of future validations, limitations or rejections, a fascinating aspect remains the novelty of the hypotheses arising from the evolutionary simulations. The sheer unexpectedness of the corresponding empirical results outside the context of evolutionary computation is an argument in favour of the claim that the quasi-neural development in these simulations can yield analogous mechanisms to neural mechanisms in biological brains.

The time costs of neural processing, moreover, were a crucial constraint in these simulation studies. With the implausible instantaneous processing assumption that is common in neural network modelling, the negative feedback loops required for the oscillatory mechanisms would not have evolved. The oscillations had strong fitness-enhancing benefits for the agents' behaviour that will probably apply to human behaviour as well. Therefore, if neural structures are chosen as the unit of selection in the simulation of the evolutionary development of a behaviour or mental function, novel, unexpected, models can emerge that are very much relevant to cognitive neuroscience.

Conclusion

Psychology – as the study of mental function and behaviours – cannot fully attain its goals without addressing the ultimate causes for the studied phenomena in addition to the study of their proximate causes. Although this assertion is accepted in evolutionary psychology, it is not yet commonplace to take neural structures as the unit of selection in the evolutionary process. Currently, if the ultimate cause for a behaviour or cognitive mechanism is investigated, genes are erroneously assumed to code directly for the trait.

In the study of the ultimate cause of behaviour it is important to recognize that not all theoretically advantageous traits are likely to have evolved. By addressing neural structures as the units of selection, and also considering their costs, it is possible to achieve more plausible insights into the solutions to problems that had to be overcome during the evolution of the brain. Conversely, models that do not build-in such constraints may evolve solutions that are wonderful in theory but implausible or highly unlikely in reality. Consider for comparison a simulation in which agents evolve a set of wheels for propulsion, or jet engines for that matter. Such traits could be highly advantageous in simulated environments, but it would be hard to see what these results could teach us about the evolution and actual behaviour of biological organisms.

Many studies investigating evolutionary behaviour focus on seemingly maladaptive behaviour. Curiously, neural constraints are almost always ignored as a possible explanation. To give an example, Thierry and Anderson listed a number of hypotheses in an extensive review for the ultimate cause of adoption, even including kidnapping, among primates.[77] They proposed kin selection may explain it as a beneficial investment in a genetically related adoptee. Alternatively, they argued that an adopter can gain experience in raising

[77] B. Thierry, J.R. Anderson, *Adoption in Anthropoid Primates*, "International Journal of Primatology" 1986, vol. 2, pp. 191–216.

young without the risk of harming its own offspring. Or it could be a form of reciprocal altruism and the adopter gains goodwill from the adoptee later in life. Or it may not be beneficial at all, but because of its rarity, natural selection failed to eliminate such behaviour. The possibility that the behaviour is indeed not beneficial by itself, yet its neural substrate is parsimonious within the constraints of neural costs, is entirely overlooked. For instance, the young females could have an evolutionary prepared urge to nurture that is so strong that, when triggered by the features of newborns, they will often even care for unrelated infants. However, when it is too costly for the adopter to have the more elaborate neural circuitry required to prevent such 'false alarms', the behaviour would constitute a solution of optimal parsimony, not a 'failure' of natural selection. The notion that similar evolutionary prepared urges could explain many kinds of altruism, moreover, is similarly neglected in almost all seminal papers on this issue.[78]

The laws of physics, including those of the brain, will always limit a species' evolutionary possibilities. Boundaries in space, time, processing capacity and energy consumption, always impose constraints on neural solutions. We argue here that incorporating these neural constraints into evolutionary models will strongly improve psychological accounts of human behaviour.

---

[78] E.g. R. Trivers, *The Evolution of Reciprocal Altruism*, "Quarterly Review of Biology" 1972, vol. 46, pp. 35–37; M.A. Nowak, C.E. Tarnita, E.O. Wilson, *The Evolution of Eusociality*, "Nature" 2010, vol. 466, pp. 1057–1062.

Wojciech Załuski
Jagiellonian University
Copernicus Center for Interdisciplinary Studies

On (Un)Happiness from the Viewpoint of Evolutionary Theory and Neuroscience[1]

Introduction

In this essay we shall present the main insights into the problem of happiness that have been provided by evolutionary theory and neuroscience and reflect on the limits of the scientific approach to this problem. We shall start by distinguishing arguably the three main questions about happiness, namely the *conceptual*: what is happiness; the *empirical*: what are the conditions of happiness; and the *axiological*: can happiness be an ethical ideal. As it transpires, rather unsurprisingly, evolutionary theory and neuroscience can be helpful in tackling the empirical question and unhelpful in tackling the remaining two questions. We shall also see that evolutionary theory inspires some interesting, though rather speculative, hypotheses regarding a more general problem (which can be viewed as connected with the empirical question) of whether human beings are "destined" to be happy.

[1] This paper was written within the research grant "The Limits of Scientific Explanation" sponsored by the John Templeton Foundation.

1. The Conceptual Question

In the history of philosophical thought two main concepts of happiness were developed: the eudaimonic and the hedonic (psychological). Both concepts can be treated as explications of a general account of happiness as a high quality of life.

Hedonic happiness is psychological welfare. This concept of happiness implies, first, that in order to determine whether an agent is happy or not, one must draw up a balance of her personal pleasant and painful experiences, and, second, that each person is the best judge of her own happiness. Briefly: *hedonic happiness is feeling good.* To make this concept of happiness more determinate one must specify how such a balance is supposed to be made. Two problems must be resolved in this context: (a) whether hedonic happiness should be assessed on the basis of the agent's experiences over her lifetime or on the basis of her experiences at some particular moment (period) or moments (periods) of her life; (b) how psychological welfare is to be defined. Ref. (a). One can hardly call a person happy who has experienced only one happy moment in her life. Thus, it seems plausible to maintain that each person's happiness can be assessed only by taking into account this person's experiences over her lifetime. Two additional options are open here: one can assess happiness on the basis of total pleasure (or net pleasure – see point (b) below) or average pleasure (total pleasure divided by the number of years lived). The first option seems more plausible, for, arguably, one cannot discount, while defining a person's life to be happy or not, this person's length of life. Ref. (b). There are two main ways of striking a hedonic balance. The first consists in taking into account only the pleasant moments experienced over a lifetime. Thus, a happy life would be a long life abounding with pleasant moments. The second one consists in taking into account both pleasant and painful moments and thereby making a net balance of pleasure. Thus, a happy life would be a long life in which there is a surplus of pleasure over pain, and a life would be happier, the larger this surplus is. The second way seems more plausible, for, arguably, a person's amount of negative experiences cannot be discounted in the overall "hedonic evaluation" of her life. This way may also be supplemented by taking into account the fact that the surplus may be generated by pleasures of different intensity. Now, it seems that among the pleasures of a happy person one must find intense ones, because it may be rather dubious to call a person happy if they have lived a monotonous, dull life, consisting of small pleasures but deprived of much pain and thereby producing a high surplus of pleasures. In sum: it seems that in order to decide whether a given person's life was happy, one should make a balance of this person's pleasures and pains over her lifetime, and know this person's total number of pleasures and the number of her intense pleasures. Thus, one can generally say that a given person can be called hedonically happy if her lifespan was long, the hedonic balance was positive, she experienced many pleasures, and many of these pleasures were intense. But this account is, of course, very sketchy. To flesh it out one would have to enrich it by taking into account such subtle relationships as, for example, the fact the presence of strongly negative experiences in one's life may have positive effects for the overall balance of our happiness (because such experiences are likely to increase our satisfaction with even small pleasures), or the shape of the distribution of pains and pleasures over a lifetime (arguably a life is more happy if its "pleasure curve" is increasing rather than decreasing, i.e., painful moments are concentrated in the earlier stages of life, and pleasant ones in its later stages).

Eudaimonic happiness is the realization of some normative idea of a good life, e.g., a life which consists in the possibly extensive development of human potentialities, or in living in accordance with nature. Briefly: *eudaimonic happiness is being good.* According to the most famous – Aristotelian – conception of eudaimonic happiness, human beings can be called happy if they pursue a life in accordance with reason, because reason is a *differentia specifica* of the human species. This concept of happiness implies the existence of a connection between happiness and virtue. This connection can be construed in at least three different ways: happiness is virtue; virtue is a necessary and sufficient condition of happiness; virtue is a sufficient condition of happiness. The first way seems most plausible: the remaining two are equivocal because they seem to assume the hedonic understanding of happiness and tacitly identify eudaimonic happiness with virtue. The concept of eudaimonic happiness implies that human beings may not know what is good for them, as they may not know their own potentialities (or, more generally, they may ignore a normative conception of a good life): there are pleasures which hinder the development of human potential and pains which promote this development. Consequently, one may feel hedonically happy without being eudaimonically happy and one may be eudaimonically happy without feeling hedonically happy (and even without knowing that one is eudaimonically happy).

There exist no conceptual relations between these two concepts of happiness, but these concepts may remain in empirical relations, e.g., eudaimonic happiness may conduce to hedonic happiness (the existence of such a relation was posited especially in classical – ancient and medieval – philosophy).[2]

2. The Empirical Question

In asking the question about the conditions of happiness, we shall mean by "happiness" its hedonic variety, because it is clear that the conditions of eudaimonic happiness are determined by an assumed conception of a good life. The question about the conditions of hedonic happiness can be posed in two different, though interrelated, ways: (1) is an individual's level of happiness constant or changeable?; (2) does the level of our happiness depend on us?

---

[2] More on the conceptual question can be found, for example, in R. Nozick, *The Examined Life: Philosophical Meditations*, Simon & Schuster, New York 1990, pp. 99–117 and W. Tatarkiewicz, *O szczęściu*, PWN, Warszawa 1979, pp. 15–29.

Ref. (1): one can formulate three main hypotheses in response to this question.

(a) *Hypothesis of a constant level of happiness.*

According to this hypothesis, happiness is entirely determined by inner factors: our biological constitution ("fate"). This constitution decides how we deal with stress, whether we have an easygoing personality, what our personal level of anxiety and stress is, etc.

(b) *Hypothesis of a radically changeable level of happiness.*

According to this hypothesis, happiness is entirely determined by events beyond our control ("chance"), or is entirely determined by our conscious efforts, or is entirely determined both by our conscious efforts and events beyond our control. This hypothesis has therefore three variants.

(c) *Hypothesis of a moderately changeable level of happiness.*

According to this hypothesis, happiness is entirely determined by our biological constitution as well as by our conscious efforts and/or by events beyond our control. This hypothesis therefore also has three variants.

Ref. (2): one can formulate three main hypotheses in response to this question.

(d) *Hypothesis of a total independence of happiness of our conscious efforts.*

According to this hypothesis, happiness depends solely on our biological constitution and/or events beyond our control. This hypothesis has therefore three variants.

(e) *Hypothesis of full dependence of happiness on our conscious efforts.*

According to this hypothesis, happiness depends solely on our conscious efforts.

(f) *Hypothesis of a partial dependence of happiness on our conscious efforts.*

According to this hypothesis, happiness depends on our conscious efforts and on our biological constitution and/or events beyond our control. This hypothesis has therefore three variants.

Different philosophers have assumed different hypotheses. For instance, the Stoics believed in hypotheses (b) and (e): they therefore believed that "fate" and "chance" play no role in determining the level of our happiness.[3] The hypotheses (c) and (f) (in their variants saying that human happiness depends on "fate", "chance", and our conscious efforts) seem to be in accordance with the commonsense account of the conditions of happiness. It should be noted, though, that there are at least two interesting arguments for the *prima facie* rather strongly paradoxical hypotheses (a) and (d). The first argument relies on the assumption of the existence of the so called "happiness attractor" (the natural – biological – level of our happiness to which we are "attracted"), which is supposed to explain the fact that even important changes in our life do not, in the long run, seem to change the level of our happiness, that is, regardless of what we experience in our life our level of happiness appears to remain more or less the same.[4] The second argument appeals to research on identical twins, which is purported to show that our level of happiness is determined by our genetic makeup.[5] Interestingly, the hypothesis (b) (in its version saying that happiness is entirely determined by events beyond our con-

[3] McMahon makes an interesting remark regarding "luck" (chance) and "fate" as factors influencing human happiness: "Strictly speaking, luck and fate are opposed, in that one implies randomness and the other pre-established order. When considered from the standpoint of human happiness, however, the two are closely related, in that each denies the role of human agency in determining the course of human events. Whether the universe is predetermined or unfolds chaotically, what *happens* to us – our happiness – is out of our hands", D.M. McMahon, *The Pursuit of Happiness. A History from the Greeks to the Present*, Penguin, London 2007, p. 10".

[4] Cf. J. Czapiński, *Szczęście – złudzenie czy konieczność? Cebulowa teoria szczęścia w świetle nowych danych empirycznych*, [in:] *Złudzenia, które pozwalają żyć*, eds. M. Kofta, T. Szustrowa, PWN, Warszawa 2001, pp. 266–306.

[5] Cf. D. Lykken, *Happiness: The Nature and Nurture of Joy and Contentment*, St. Martin's Press, London 2000. In this book, Lykken, drawing on his study of the life histories of some three thousand identical and fraternal twins, argues for the existence of a "happiness set point" largely determined by our genetic makeup (as Lykken writes, "Nearly 100 percent of the variation across people in the happiness set point seems to be due to individual differences in genetic makeup (p. 58)". He has conceded, though, that our efforts and events beyond our control play an important role in whether we succeed in achieving our "happiness set point".

trol) and the related hypothesis (d) (in its version saying that happiness depends solely on events beyond our control) can be ascribed to the ancient Greeks as their natural, pre-philosophical attitude towards happiness. This attitude is reflected in the Greek term for happiness – *eutychia*, which means "good luck"; thus, according to ancient Greeks, happiness is "a matter of luck".[6]

In the remainder of this section we shall evaluate the extent to which biological sciences – especially evolutionary theory and neuroscience – can contribute to answering the empirical question, that is, can help decide between the competing hypotheses about the conditions of happiness.

(*Evolutionary theory*) Evolutionary theory does not say anything about individual – personality-based – differences in the levels of happiness, i.e., it suggests that our biological constitution is more or less the same, but it seems to lead to three general hypotheses about happiness *in general*, i.e., about what this uniform biological constitution is like. *First*, from the standpoint of evolutionary theory, happiness is not an ultimate goal of human actions but a means to realizing a fundamental evolutionary goal: survival and reproduction (and therefore spreading our genes). Humans are therefore not "destined" to be happy but, rather, are "destined" to effectively pass on their

---

[6] Cf. D.M. McMahon, *The Pursuit...*, *op. cit.*, p. 68. In point of fact, happiness was linked with luck and fate for the larger part of human history; the above-mentioned pre--philosophical attitude of ancient Greeks was therefore also a pre-philosophical attitude of other peoples. As Darrin McMahon wrote: "It is striking that in virtually every Indo-European language, the modern word for happiness is cognate with luck, fortune, or fate. The root of "happiness", for example, is the Middle English and Old Norse *happ*, meaning chance, fortune, what *happens* in the world, giving us such words as "happenstance", 'haphazard", "hapless", and "perhaps". The French *Bonheur*, similarly, derives from *bon* (good) and the Old French *heur* (fortune or luck), an etymology that is perfectly consistent with the Middle German *Glück*, still the German word for happiness *and* luck. In Italian, Spanish, and Portuguese, *felicità*, *felicidad*, and *felicidade* all stem from the Latin *felix* (luck, sometimes fate), and the Greek *eudaimonia* brings together good fortune and good god. (...) In the Indo-European language families, happiness has deep roots in the soil of chance (pp.10–11)". Thus, the primitive meaning of "eudaimonia" had a similar meaning to the "eutychia"; only later philosophers gave it a new meaning – of a "good life".

genes to subsequent generations. They are "designed" to experience happiness only so long as it serves the evolutionary goals. Happiness accompanies activities which promote these evolutionary goals (e.g., eating, sex, pursuing social esteem) and thereby "encourages" human beings to undertake these activities. Such activities trigger off neurobiological mechanisms which give rise to the feelings of pleasure (i.e., trigger off the reward mechanisms) and the expectation of these feelings moves us to undertake these activities. Thus, evolutionary theory points at the same time at the main types of the sources of our happiness: these sources are activities promoting fundamental evolutionary goals. It should be stressed, though, that according to the adaptive lag-hypothesis (saying that our biological adaptations which were adaptive in ancestral environments in which our biological and psychological dispositions had been shaped may fail to be adaptive in modern environments) our biological system of reward (dopaminergic system) may "misfire", i.e., pleasure (hedonic happiness) may accompany non-adaptive activities. In other words, the axis "pleasure-pain" may diverge from the axis of "biological fitness – the lack of biological fitness". *Second*, in order for happiness to play its evolutionary role (i.e., to motivate to undertake fitness-enhancing activities), it must be ephemeral and transitory (because durable happiness would amount to the lack of evolutionary progress) as well as not excessively euphoric (because durable happiness would probably hinder the proper performance of our daily activities, which are a constant and self-propelling oscillation between satisfaction and dissatisfaction). But evolutionary theory suggests also that even though happiness is transitory and non-euphoric human beings should be expected to retain hope that they will reach a durable and euphoric happiness, since such hope is a powerful protection against their falling into despair and, consequently, against what may be called an "evolutionary impasse". Thus, human beings can be expected to cherish an illusionary hope that they can achieve durable happiness, since without such a hope they would not so readily engage in activities leading to a transitory variety of happiness – necessary for the realization of the evolutionary goals. The phenomenon which consists in that human beings are "designed" to become involved in a rather hopeless and painful pursuit of durable happiness is referred to as "hedonic treadmill"; Darrin McMahon describes this phenomenon in the following way:

> human beings display an undeniable tendency to quickly accommodate themselves to their pleasures – to grow bored – and then become anxious or uneasy in their satisfaction. Like junkies in need of a fix, we need a variety in our pleasures or greater doses of the same when initial "rush" wears off. It is this longing – a form of pain – that sends un in renewed pursuit. (...) the hedonic treadmill (...) has been adapted by natural selection to sustain us in life. (...) although natural selection has conditioned us to experience pleasure in activities that contribute tour survival, it disposes us to quickly adapt to them and then to strive for a little more. In this view, enduring satisfaction or permanent contentment would not be conducive to survival. It is in our interest – and so in our genes – always to be slightly wanting, restlessly searching for further satisfaction. A bit if anxiety keeps us on guard against danger, and a bit of unfulfilled desire keeps us on the chase, ever eager to ensure our survival and that of our kin.[7]

It should be noted that this phenomenon was well known to many thinkers and religious reformers (e.g. Schopenhauer and Buddha) and constituted one, perhaps especially important, grounds of their pessimistic evaluation of the human condition. The merit of evolutionary theory lies, therefore, not in discovering this phenomenon but in explaining its origins. *Third*, evolutionary theory suggests that, in general, human beings are rather unhappy in the deepest layer of their personality, although hopeful of attaining happiness; they are unhappy in the following sense: since their ancestors lived in a hostile world in which things that posed a danger to them were more numerous than

[7] D.M. McMahon, *The Pursuit...*, *op. cit.*, pp. 422–423.

those which favoured them (or perhaps more precisely: in which non--avoiding dangerous things was much more costly than overlooking favourable things) they must have developed acute vigilance in order to avoid those dangerous things. Thus, a "mental climate" which prevails in human lives is the climate of anxiety, of worry arising out of an awareness of present and future dangers. In other words, for the larger part of their life human beings experience dejection and worry, because their brains reached their final form at the time when fear and alertness were adaptive emotions. One can therefore say that evolutionary theory supports Heidegger's claim that the fundamental category of human existence is *Sorge* (care). This claim is consistent with a well-justified hypothesis according to which human beings exhibit the "negativity bias": their brains are constructed above all with a view to noticing and reacting against dangers and obstacles; accordingly, their reactions and feelings towards dangerous things are faster and more intense than reactions and feelings towards positive things.[8] The assertion that, from the standpoint of evolutionary theory, human beings are in the deepest layer of their personality rather unhappy, is of course also fully consonant with and supported by the aforementioned claim about the existence of the hedonic treadmill. *Fourth*, evolutionary thinking is an inspiration of the following, assuredly apt, remark of Bertrand Russell about unhappiness arising from the easy satisfaction of natural needs: "The human animal, like others, is adapted to a certain amount of struggle for life, and when by means of great wealth *homo sapiens* can gratify all his whims without effort, the mere absence of effort from his life removes an essential ingredient of happiness".[9] The aforementioned evolutionary hypotheses about human happiness seem very plausible, with the possible excep-

---

[8] On negativity bias see, e.g., J. Haidt, *Szczęście. Od mądrości starożytnych po koncepcje współczesne*, translated by A. Nowak, GWP, Gdańsk 2006, pp. 62–67. On the adaptive value of low moods see, e.g., R.M. Nesse, *Is Depression an Adaptation?*, "Arch Gen Psychiatry" 2000; vol. 57, pp. 14–20.

[9] B. Russell, *The Conquest of Happiness*, W.W. Norton & Company, Inc., New York–London 1996 (First published in 1930).

tion of the hypothesis that human beings are, at the deepest layer of their personality, rather unhappy: this last hypothesis is likely to be at odds with some or perhaps even many people's intuitive sentiment that the natural state of human beings is that of "hedonic indifference" or even moderate happiness rather than *Sorge*.

(*Neuroscience*) Neuroscience provides some insights not only into human happiness in general but also into the causes of individual differences in the levels of happiness: it sheds some light on what biological factors decide that some human beings are, over their lifetime, happier than others. Let us make a short survey of this research. *First*, as is well known, research on the brain has shown that the human brain is composed of various parts fulfilling different, heterogenic functions. There are two main divisions of the human brain. The first one is between the "old (emotional) brain" (the limbic system: hypothalamus, hippocampus, and amygdala) and the "new (intellectual) brain"; the other is between left hemisphere and right hemisphere. Jonathan Haidt plausibly argues that it is doubtful that agents equipped with such a brain – with a "divided self"– may ever attain happiness.[10] This insight harmonizes with the above mentioned evolutionary insight that, to use the Heideggerian jargon, the central category characterizing human existence is the category of *Sorge*. Clearly, such a structure of the brain is fully understandable from the standpoint of evolutionary theory: the human brain has been built step by step over an enormously long period of time in the process of natural selection rather than having been planned from the start by some intelligent designer. *Second*, the neurobiological research has demonstrated that when human beings experience hedonic happiness their left prefrontal lobe becomes activated. This brain structure comprises many neurotransmitters (dopamine, serotonin, GABA) which are responsible for the transport of positive signals to the limbic system. Research on one-year children has shown that those of them who have more active left prefrontal lobe do not cry when left alone as opposed

[10] Cf. J. Haidt , *Szczęście...*, *op. cit.*, pp. 17–52.

to children with a less active left prefrontal lobe. Accordingly, it has been hypothesized that this brain structure is responsible for our affective style and thereby for our "(un)happiness potential"; if one's right prefrontal lobe is more active than the left one then one exhibits the style which consists in distancing, worrying that one may make a mistake, preoccupation with potential criticism – in short: one is then "predestined" to be more unhappy. One can therefore say that, to a certain extent, some human beings are genetically predestined to be more happy than others (or less unhappy than others) – "to a certain extent" because the human brain is plastic, especially before the period of maturity, and various factors (e.g., intense stress) may make a child less happy than would result from her genetic makeup.[11] This style is therefore not unchangeable: one may change to a certain degree by, for example, meditation, cognitive therapy, or various medicines. As for the last way of changing our affective style, it should be mentioned that research conducted within psychopharmacology and neuroscience has shown that – and how (i.e., through the medium of what biological mechanisms) – certain medicines may influence our hedonic happiness (e.g. Prozac may change our affective style by making us worry less – by making us perceive the world as full of positive possibilities rather than as a source of potential dangers).[12] *Third*, Hans Eysenck distinguished three dimensions of human personality (introversion-extraversion; neuroticism; psychoticism and argued (*inter alia*) that these three dimensions have neurobiological bases.[13] Now, it has been hypothesized that our individual level of happiness is determined by a combination of these dimensions in such a way that happy people are most frequently extrovert, non-neurotic, and non-psychotic.

---

[11] But one may, of course, ask whether this brain structure creates or reflects the state of happiness.

[12] Cf. P.D. Kramer, *Listening to Prozac*, Penguin Books, Toronto 1994.

[13] Cf. H.J. Eysenck, *Dimensions of Personality*, Transaction Publishers, London 1997 (First published in 1947); H.J. Eysenck, S.B.G. Eysenck, *Psychoticism as a Dimension of Personality*, Hodder and Stoughton, London 1976.

Let us summarize. The research pursued within the biological sciences has led to two main conclusions related to the hypotheses about the conditions of human happiness. The first conclusion (supported especially by neuroscience) says that much of variance in the levels of happiness experienced by human beings can be accounted for by variance in our biological constitutions. Consequently, this conclusion undermines those hypotheses about the conditions of human happiness which neglect the role of "fate" – of our biological constitution – in determining the levels of our happiness (i.e., those hypotheses which explain individual differences in human happiness only by events beyond our control and/or our conscious efforts). The second (more controversial) conclusion concerns not so much the conditions of human happiness, as human happiness in general: it asserts that whatever conditions may determine our level of happiness, we shall always be, in the deepest layer of our personality, rather unhappy than happy. To reconcile these two conclusions one can simply reformulate the former by saying that those with a "favourable" biological constitution will experience over their lifetime less unhappiness than those with an "unfavourable" biological constitution.

3. The Axiological Question

The axiological question will be answered differently depending on the assumed concept of happiness. Clearly, eudaimonic happiness ("being good") can be an ethical ideal – and as such was assumed within classical philosophy. What is controversial is whether hedonic happiness ("feeling good") can be regarded as an ethical ideal. Hedonistic ethics assumes that hedonic happiness is the highest value, the highest good, or an intrinsic value. Anti-hedonistic ethics assumes that hedonic happiness cannot be an ethical ideal. Anti-hedonistic ethics may take a more or less negative attitude towards hedonic happiness. According to asceticism, hedonic happiness is inherently wrong. According to other anti-hedonic ethics (e.g. Platonic) hedonic happiness is not inherently wrong but it is wrong when it is given priority over other – higher – values (as was memorably put by Kant, one of the most famous representatives of the anti-hedonistic ethics, we should not aim at being happy but at becoming worthy of happiness). There are several plausible ethical arguments in favour of anti-hedonistic ethics, i.e., arguments against treating hedonic happiness as an intrinsic value: we have qualms of conscience when in the case of conflict between hedonic happiness and other goods we have given priority to the former; we give a higher valuation to hedonic happiness if it is generated by realization of some praiseworthy deeds; hedonic happiness flowing from evil acts is not regarded as good.[14] These arguments show that hedonistic ethics is at odds with our deeply embedded ethical intuitions. One can also formulate non-ethical arguments against treating hedonic happiness as an ethical idea. For instance, it has been maintained that hedonic happiness would be a self-defeating ideal: it is not attained by those who consciously pursue it; it can only be attained as a "side-effect" of our pursuing other valuable ends.[15] Another non-ethical or perhaps quasi-ethical argument against treating hedonic happiness as an intrinsic value (an argument advanced by many thinkers, especially by Nietzsche), says that hedonic happiness, unlike physical and mental suffering, makes us uncreative, hinders us from discovering our potentialities. As we can see, in answering the axiological question it is redundant to appeal to neuroscience and evolutionary theory: science cannot help in any interesting way to tackle the fundamental ethical problem of the value of hedonic happiness.

[14] Cf. W. Tatarkiewicz, *O szczęściu*, *op. cit.*, pp. 522–523.

[15] Cf. V.E. Frankl, *Paradoxien des Glücks. Am Modell der Sexualneurose*, [in:] *Was ist Glück?* Ein Symposion, Dtv-Taschenbücher 1134, Dtv-Verlag, München 1976.

Conclusion

The above considerations show that the biological sciences (evolutionary theory and neuroscience) can provide some non-trivial insights into the empirical question about human happiness. Especially interesting contributions seem to be the evolutionary claim about the necessarily transient nature of happiness and the generally unhappy nature of human beings and the neuroscientific insight regarding the individual – happiness-relevant – differences in our biological constitution. But, undoubtedly, there is still much work be done within the biological research on happiness. In particular, the challenge for this research is to test some more subtle empirical hypotheses about happiness. Let us, at the end of this essay, provide some examples of such hypotheses: only hedonically happy persons are good persons, i.e., only hedonically happy persons can exhibit disinterested concern with the well-being of other persons; the fewer contradictory desires one has, the more one is likely to attain hedonic happiness; hedonic happiness eludes us if we intentionally pursue it;[16] what human beings really aim at is not experiencing hedonic happiness, but attaining the basis of hedonic happiness;[17] only a person who is unhappy in her innermost core desires hedonic happiness;[18] eudaimonic happiness is a necessary (in the empirical sense) condition of hedonic happiness, i.e., only good people can "really" feel good; human beings do not always pursue hedonic happiness (this last hypothesis is especially interesting, because confirming it would amount to undermining psychological hedonism – a theory of human motivation which says that pleasure seeking and pain avoidance are the only motives of our actions).

---

[16] Victor Frankl (*Paradoxien..., op. cit.*) calls it "the first paradox of happiness" ("the second paradox of happiness" analysed by Frankl consists in that even in the midst of the most terrible events, even in extreme suffering (e.g., in a concentration camp) one may find happiness if one is capable of giving sense to these events).

[17] Cf. V.E. Frankl, *Paradoxien..., op. cit.*

[18] This claim was defended by Max Scheler (cf. M. Scheler, *Cierpienie, śmierć, dalsze życie*, translated by A. Węgrzecki, PWN, Warszawa 1994).